Teaching Physics through Ancient Chinese Science and Technology

中国古代科学技术与物理教学

Teaching Physics through Ancient Chinese Science and Technology

中国古代科学技术与物理教学

Matt Marone
Mercer University, Macon, Georgia, USA

Morgan & Claypool Publishers

ISBN 978-1-64327-458-4 (ebook)
ISBN 978-1-64327-455-3 (print)
ISBN 978-1-64327-456-0 (mobi)

DOI 10.1088/2053-2571/ab03cb

Version: 20190701

IOP Concise Physics
ISSN 2053-2571 (online)
ISSN 2054-7307 (print)

A Morgan & Claypool publication as part of IOP Concise Physics
Published by Morgan & Claypool Publishers, 1210 Fifth Avenue, Suite 250, San Rafael, CA, 94901, USA

IOP Publishing, Temple Circus, Temple Way, Bristol BS1 6HG, UK

To polymaths the world over.

Contents

Cover ix

Preface xi

Author's note xiii

Acknowledgments xiv

Author biography xv

1 Introduction **1-1**

Notes on the Chinese Language and Terms 1-6

Is it Really Science? 1-9

References 1-16

2 Kinematics (运动学 Yùndòngxué) **2-1**

References 2-14

3 Force (力 lì) **3-1**

References 3-12

4 Torque (力矩 lìjǔ) **4-1**

Reference 4-4

5 Energy (能 Néng) and Momentum (动量 Dòngliàng) **5-1**

Energy Conservation (能量守恒 Néngliàngshǒuhéng) 5-4

Momentum and Force 5-7

References 5-10

6 Experiments **6-1**

List of laboratory experiments 6-2

6.1 Paper making (造纸 Zàozhǐ) 6-3

Objective 6-3

Chinese vocabulary 6-3

Materials 6-3

Introduction 6-3

Experimental 6-5

Paper making part 2 6-9
Analytical method 6-9
Varying the drying force 6-9
Procedure for paper making, data sheet—force variation 6-11
Measuring the thickness 6-11
Analysis 6-12
Points to consider in analysis 6-17
Alternative methods of analysis 6-18
Data sheets: For a paper-making experiment 6-20
6.2 Stress–strain curves of silk thread (丝线应力应变曲线 Sīxiàn yìnglì-yìngbiàn qūxiàn) 6-23
Objective 6-23
Chinese vocabulary 6-23
Materials 6-23
Introduction 6-23
Experimental 6-25
Procedure to degum the cocoon 6-25
Silk reeling 6-25
Part 2: Stress–strain testing 6-27
Stress–strain measurements 6-29
Procedure for stress–strain analysis 6-30
Analysis 6-31
Methods for measuring the diameter of the silk strand 6-32
Data sheets: For silk stress–strain experiments 6-37
6.3 Steelyard balance (秤 Chèng) 6-39
Objective 6-39
Chinese vocabulary 6-39
Materials 6-39
Introduction 6-39
Experimental: construction 6-40
How to use the 秤 chèng balance 6-45
Analysis 6-47
Data sheets: For 秤 Chèng steelyard balance experiments 6-51
References 6-54

Cover

The object on the title page and on the cover of the print edition is a steelyard clepsydra or water clock 'rediscovered' in 2012 by the author at the Needham Research Institute (NRI) at the University of Cambridge. Thus far, no published information concerning this timekeeping device has been found in the literature. The only clue to its history can be found in a footnote in *The Hall of Heavenly Records*, p 84, which is reproduced below.

> [87] The statement has often been repeated (e.g. in Jeon, STK, p 148; and more cautiously, D R Hill *Arabic Water-Clocks*, p 5) that a siphon used with a clepsydra outflow vessel of the non-constant level type can partly compensate for the elects of unequal and declining water pressure. This if true would make the use of a siphon advantageous as compared with the use of a simple outflow tube located at the bottom of the tank.
>
> In a series of experiments kindly conducted for us by Sir Brian Pippard of the Cavendish Laboratory, Cambridge University, this alleged effect did not manifest itself. The outflow rates for both siphons and simple outflow tubes were found to be nearly the same (within 1 per cent) under a variety of conditions. A siphon can compensate for unequal and declining pressure in the

> outflow vessel only if the siphon itself is carried on a float within the vessel. We know of no East Asian clepsydra that include any such arrangement.

The author reassembled the device from parts in the NRI archive and has conducted some experiments to establish volume flow rates with and without the siphon. It has become emblematic of our method of exploring the underlying physics of ancient Chinese technology.

Reference

Needham J, Lu G and Major J S 2004 *The Hall of Heavenly Records: Korean Astronomical Instruments and Clocks 1380–1780* (Cambridge: Cambridge University Press)

Preface

You may wonder how this book and the related class came about. My interest in all things relating to China began when I was young. In the early 1970s I became a Ham radio operator. Before the Internet, we had actual paper maps and since I was operating a shortwave radio station, my room had quite a few maps on the walls. The largest and most conspicuous map was one of China. I used to marvel at the strange place names and diverse topography of that mysterious country. How did one pronounce those strange place names? This was also about the time that US–China relations were starting to improve and China was in the news. Why was the capital called Peking, Beijing, Pei Ping? So many names for the same place, that was strange. China's enigmatic leaders, blue Mao jackets, and huge population were images that captured my interest. For some reason they were disavowing a culture that has lasted for thousands of years and made them one of the oldest continuous civilizations on the planet. Once, as I was operating my shortwave transmitter, looking for someone to communicate with in Morse code, I received a few brief characters from a very weak signal. The call sign was that of a Chinese radio station. China? I thought. Given the situation in the country at the time, how could that be? Unfortunately, the ionosphere had its way with my attempted communication, and the signal faded out. I was fascinated with the country and its people. When my birthday came around, I often went to a Chinese restaurant to celebrate. I read books about China, and its mythology and philosophy. Years later, as a college student and in graduate school, I had the opportunity to meet Chinese students and learn more about their culture. We ate together, celebrated holidays, attended Chinese language church services and I grew to appreciate their way of life. Then, the ultimate learning experience was offered to me. A fellow graduate student working in the same research group helped me to contact her former professor at the Chinese Academy of Sciences. While many Chinese students were on their way to the United States, I was on my way to work as a post-doctoral researcher at the Chinese Academy of Sciences in Beijing. As a foreigner in China I could not exactly blend in, but I did what I could to live and work like the people in my community. I now had firsthand access to scholars and museums. Of all the amazing experiences I had in China, none compare with that of meeting my wife. The development of my class on Chinese science and the publication of this text are all the direct result of her inspiration and encouragement. Like all Chinese, she is proud of her country's heritage and scientific accomplishments. Most people I spoke to in China were proud of their long history and aware of the four great inventions (四大发明, sìdàfā míng). These inventions are paper, printing, the magnetic compass and gunpowder. They are aware of these inventions in a largely historical sense, in much the same way that an American is aware of some general facts about George Washington or Abraham Lincoln. Each of these inventions contains, perhaps hiding in the background, a great deal of science and engineering. In this text we shall consider the underlying science, historical, cultural and occasionally metaphysical connections related to several ancient Chinese inventions. As an applied physicist I see great value in learning science by recreating ancient inventions and technology. Thus, this text contains experimental sections suitable for laboratory instruction.

In 2012 I had the great privilege to visit the Needham Research Institute (NRI) at the University of Cambridge. The idiom 'like a child in a candy shop' is the only fitting way to describe my experience. Here I was surrounded by all of Joseph Needham's reference material, notes and artifacts. The staff at the NRI were hospitable and gracious and gave me the freedom to explore their collections.

The material I describe here is based on an introductory physics class taught at Mercer University. This class, 'Ancient Chinese Science and Technology', is aimed at non-science majors who must fulfill the general education laboratory science requirement. The syllabus for this class may be found on the Mercer University Physics Department web page (Marone 2016). A unique aspect of this book is the section on laboratory experiments, all based on ancient Chinese technologies. Detailed laboratory procedures for each experiment are included. We hope that this text will be useful for STEM and high school science instruction as well general education students. Educators in blended language classrooms should find this work a useful reference. As you will see, this text combines aspects of Chinese science, language, history and philosophy in what we believe is a unique and instructive blend.

Reference

Marone M 2016 *Ancient Chinese Science and Technology* http://physics.mercer.edu/curriculum/syllabi/phy108marone(f16).pdf

Author's note

The material presented in this book is not intended to be an exhaustive treatment of all topics in physics. It is very specifically aimed at non-science majors and covers only a few areas commonly found in a two-semester introductory physics sequence. The topics selected reflect the author's own research areas and the structure of the class taught at Mercer University. These topics were also chosen based on the natural connection between their early development in China and how they can serve as examples of particular physical principles. Many of the laboratory experiments are related to building some sort of device. They are intended to be engaging and low cost. They require the use of common hand tools, and basic laboratory equipment. Students often tell me that they enjoy the hands-on nature of these experiments. The artifacts that students create give them a certain sense of pride and accomplishment not found in more traditional 'cookbook' laboratory exercises. Each experiment can be thought of as a nucleus or seed that the instructor can build upon to suit the needs of their particular class. Throughout this book, and especially in the laboratory component, there is always a multilayer set of learning objectives. At first it may seem that we are just recreating a particular technology or device. On another level, we are attempting to see how the ancient Chinese understood that device or technology in their own cultural context. Going further, we then investigate the underlying physical principles as we teach them in a modern science classroom. Along the way we will meet ideas and world views dramatically different from what is found in the typical Western science setting.

Acknowledgments

A work as complex as this, is dependent on the goodwill and encouragement of many individuals. When I first traveled to China, Professor Zahang Dian-Lin of the Chinese Academy of Sciences, was a wonderful mentor to me. He and his graduate student Ying Feng, helped to guide me through the many complexities of Chinese culture. I am also indebted to Dr Leona Kanter and the members of the Asian Studies Development Program whose tireless efforts created the Asian Studies minor at Mercer University. Dr Lake Lambert, former Dean of the College of Liberal Arts, provided financial support and the freedom to pursue this work. The staff of the Needham Research Institute (NRI) at the University of Cambridge were kind enough to allow access to the many artifacts and notes of Joseph Needham. Mr John Moffett, Ms Susan Bennett and Professor Christopher Cullen of the NRI provided encouragement and assistance. Dr Ronnie Littlejohn, Professor of Philosophy and Director of Asian Studies at Belmont University, has been a constant source of encouragement and supporter of blending the study of Physics and Asian Studies. Finally, I must thank my family for their support during this project.

Author biography

Matt Marone

Matt Marone is an Associate Professor of Physics at Mercer University in Macon, Georgia, where he teaches Physics, Astronomy and Asian Studies. He received his PhD and MS degrees from Clemson University in the area of experimental solid state physics and a BS degree in physics from the Rochester Institute of Technology. In the early 1990s he worked as a post-doctoral researcher at the Chinese Academy of Sciences in Beijing. At Mercer he teaches a wide range of physics classes including several specialized classes in observational astronomy, acoustical foundations of music and ancient Chinese science. In addition to his academic research, he also serves as a consultant to NASA's Marshall Space Flight Center in the area of space resources. His research in space resources is involved with the extraction of oxygen and metals from the regolith of the Moon, Mars and asteroids. As a member of the Society for Georgia Archeology, he assists in the study and interpretation of Georgia's historic and prehistoric archeological heritage.

Teaching Physics through Ancient Chinese Science and Technology

中国古代科学技术与物理教学

Matt Marone

Chapter 1

Introduction

Many readers of this text will find it difficult to understand that science and technology were developed in other areas of the world outside of Europe. To properly appreciate the message of this text, one must try to free one's mind from Eurocentric thinking and be open to the idea that other cultures developed their own models and descriptions of the natural world. Francis Bacon is considered by some to be one of the earliest advocates of what we now call the 'scientific method'. He stressed the importance of observation and inductive reasoning. Bacon made a comment that ties in directly with the study of ancient Chinese science. Consider the following quote from his famous work *Novum Organum* (Bacon 1960):

> Again, it is well to observe the force and virtue and consequences of discoveries, and these are to be seen nowhere more conspicuously than in those three which were unknown to the ancients, and of which the origin, though recent, is obscure and inglorious; namely, printing, gunpowder, and the magnet. For these three have changed the whole face and state of things throughout the world; the first in literature, the second in warfare, the third in navigation; whence have followed innumerable changes, insomuch that no empire, no sect, no star seems to have exerted greater power and influence in human affairs than these mechanical discoveries.

Perhaps Bacon should have added paper to the list. If he had, then he would have been quoting what the Chinese call their four great inventions (四大发明, sìdàfāmíng). Two of these inventions, paper and the magnet, figure prominently in the laboratory component of this class. Making gunpowder is not something on the approved list of classroom activities. Notice the Chinese terms used here. Throughout this text, authentic Chinese terminology will be used. The Chinese characters are in the simplified form adopted by the People's Republic of China (PRC).

doi:10.1088/2053-2571/ab03cbch1

The PRC began character simplification in 1956. Adjacent to the Chinese characters is a type of phonetic pronunciation guide known as Hanyu Pinyin Romanization. This system was first published in 1958 and has been revised several times.

Chinese science was largely unknown or ignored in the Western world for centuries. Shown below is a quotation from the 1901 edition of *The Middle Kingdom* by S Wells Williams, page 65 (Williams 1901). Williams adopted the Chinese name 衛三畏 (卫三畏 Wèi Sān Wèi).

> That enlargement of the mind which results from the collection and investigation of facts, or from extensive reading of books on whose statements reliance can be placed, and which leads to the cultivation of knowledge for its own sake, has no existence in China. Sir John Davis justly observes that the Chinese 'set no value on abstract science, apart from some obvious and immediate end of utility;' and he properly compares the actual state of the sciences among them with their condition in Europe previous to the adoption of the inductive mode of investigation. Even their few theories in explanation of the mysteries of nature are devoid of all fancy to make amends for want of fact and experiment, so that in reading them we are neither amused by their imagination nor instructed by their research. Perhaps the rapid advances made by Europeans, during the two past centuries, in the investigation of nature in all her departments and powers, has made us somewhat impatient of such a parade of nonsense as Chinese books exhibit. In addition to the general inferiority of Chinese mind to European in genius and imagination, it has moreover been hampered by a language the most tedious and meagre of all tongues, and wearied with a literature abounding in tiresome repetitions and unsatisfactory theories. Under these conditions, science, whether mathematical, physical, or natural, has made few advances during the last few centuries, and is now awaiting a new impulse from abroad in all its departments.

Perhaps even more startling than his racist views is the fact that he was Professor of Chinese Language and Literature at Yale. In 1877, he returned to the United States and was appointed by Yale University to teach Chinese. For this reason, Yale has been credited as the first American institution to offer courses in Chinese (http://ceas.yale.edu/about-ceas/samuel-wells-williams).

Even more astonishing is that he lived in China for some forty years. His description of Chinese art, music and literature are, as one might guess, even less flattering. The following is an excerpt from the 1871 edition of *The Middle Kingdom*, page 173 (Williams 1871).

> The deficiencies of the Chinese in music will not lead us to expect much from them in painting or sculpture, for all seem to flow so much from the same general perception of the beautiful in sound and form, that where one is deficient, all are likely to be unappreciated. This want in Chinese mind, for we are hardly at liberty to call it a defect, is, to a greater or less degree, observable

in all the races of Eastern Asia, none of whom exhibit a high appreciation of the beautiful or sublime in nature or art, or have produced much which proves that their true principles were ever understood. Painting is rather behind sculpture, but neither can be said to have advanced beyond rude imitations of nature.

My readers may feel insulted or contaminated by the inclusion of these comments but there is a reason for it and it sets the stage to understand the work of more modern scholarship. One of my most effective teaching methods is the use of contrasts and counterintuitive examples. The best way to purge your mind of these reprehensible comments by Professor Williams is to seek out a museum of Asian art or attend a Chinese musical performance.

What can we learn from Williams that will be helpful to us in our understanding of Chinese science? In his works, there is the constant drumbeat of comparison between China as he saw it in the 19th century and what he perceived as the superiority of Western civilization, most prominently displayed by Great Brittan. We must try not to fall into the same trap and, as much as it is possible, allow China and the Chinese culture to speak for itself. Of course, since the culture is so different from that of the West, we may need some points of explanation and interpretation. For example, the Chinese concept of the Dao (道 Dào) had a major influence on early Chinese science. This concept, which Williams mistakenly translates as 'reason', may be foreign to many of my readers. Thus, some words of explanation may occasionally be required. One should adopt the famed objectivity of the scientific method and extricate oneself from the mire of cultural comparisons.

Perhaps this is a good place to introduce the term *Sinology*. People unfamiliar with the term often think that it is the study of sin. In fact, the word is derived from the Latin word *Sinae*. The term is thought to have come from the Arabic word *Sin* which may have been a direct reference to the Chinese word Qin. The Qin here refers to the Qin Dynasty (秦朝, Qín Cháo), which was the first imperial dynasty of China and spanned the period 221 to 206 BC. Sometimes we use the term 'Sino' as in 'Sino-American' to indicated Chinese-American. The word 'Sino' is related to the term *Sinae*.

A quick search on the Internet will reveal the names of hundreds of sinologists. Their scholarship has influenced this work in many ways but there is one who has influenced my work more than any other. Joseph Needham and his monumental book series *Science and Civilisation in China* (SCC) played an important role in the development of the course described in this book. In 2012, I had the privilege of spending part of my sabbatical at the Needham Research Institute (NRI) located on the grounds of Robinson College at Cambridge University. Much has been written about Needham and his work. For the purpose of this book, I do not plan to trace over the many colorful exploits of his life. The Needham Research Institute (http://www.nri.cam.ac.uk/index.html) has extensive information available online. Throughout this book, I will refer to his work and the questions he raised. *Science and Civilisation in China* was first published in 1954 and new editions of the series are still being written! In 2008, *Volume 5, Chemistry and Chemical*

Technology Part 11, Ferrous Metallurgy was published and in 2015, *Volume 6, Biology and Biological Technology, Part 4. Traditional Botany: An Ethnobotanical Approach* was added to the series. It is amazing that a series of books, over 60 years in the making, is still informing the world about the contributions of ancient China. Students of sinology find it difficult to believe that Needham actually researched and understood all the material in the earlier volumes that he wrote. Of course, he did have help and coauthored the series with many other scholars, but the painstaking and detailed research notes that he used to create his masterwork are all carefully archived in the fantastic library at the NRI.

Needham, also known by his Chinese name 李约瑟 Lǐ Yuēsè, started out his scientific career as a biochemist, and was particularly interested in the field of Embryology. In the mid-1930s he met three young Chinese students who came to study at Cambridge. As the story goes, one particular member of this group, 鲁桂珍 Lǔ Guìzhēn, sparked his interest in ancient Chinese science. She also sparked a romantic interest, which seemed to have been condoned by his wife. From 1943–46, Needham lived in China as a representative of the British government. This was during the Second World War and parts of China were occupied by Japan. His mission was to aid the Chinese war effort by providing laboratory equipment and scientific books to Chinese scientists. During that time, avoiding the Japanese occupation forces, he visited universities, factories and historic sites. He met with Chinese scientists and began to amass whatever information he could find about Chinese science and technology. Much of this material made it back to Cambridge and can still be found at the NRI and in the Cambridge University Library.

Beginner students often find *Science and Civilisation in China* a bit difficult to follow. In many cases Needham makes connections with other material that may seem obscure and largely unknown to anyone other than a specialist in the field. A recurring theme in the book is the timeline of discoveries and particularly the question of priority. Much discussion is spent on trying to determine who first discovered an idea or invented a technology. There is also discussion of how such discoveries may have made their way to China from other cultures and how Chinese discoveries dispersed into the rest of the world. As important as such research is to the history of science that is not the major concern of this book. My intention is to examine the underlying physics of these inventions and discoveries.

If you are looking for a Chinese equivalent on Newton's *Philosophiæ Naturalis Principia Mathematica,* or Galileo's *Discorsi e Dimostrazioni Matematiche Intorno a Due Nuove Scienze,* I am afraid that you will be sorely disappointed. Much of the material we will examine is given as observations about nature or as a description of some technology. Mathematical demonstrations such as those of Euclid were not the way that the ancient Chinese expressed their early scientific ideas. This is why people like Williams could not recognize Chinese science for what it was. When explanations are given, they are often couched in terms of the Chinese theory of Five Elements or Five Phases (五行, Wǔ Xíng).

Much of Needham's work was historical, theoretical and at times philological. He did collaborate with John H Combridge in some hands-on historical reconstruction in the area of clockwork mechanisms as described in his work *Heavenly Clockwork.*

As Needham dug deeper and deeper into Chinese science, he was nagged by a particular thought. This same thought may occur to our readers as well. Needham discovered so many different areas of science and technology in which China was ahead of Europe at the same period of time. Yet, the China that he saw in the 1940s was very out of pace with developed countries. Needham raised the question this way in his work *The Grand Titration*, page 16 (Needham 1969):

> Why did modern science, the mathematization of hypotheses about Nature, with all its implications for advanced technology, take its meteoric rise only in the West at the time of Galileo but had not developed in Chinese civilisation or Indian civilisation?

This has become known as 'The Needham Question' or alternatively 'The Needham Puzzle'. Sometimes it is cast into a slightly different form and asks about the Industrial Revolution or the Renaissance. But the questions are more or less the same. The question itself is very controversial. Some imply that it is not even a legitimate question to ask. Noted sinologist Nathan Sivin, whose writings also influenced my own work, shares the following comments on the Needham Question (Sivin 1995):

> It is striking that this question—Why didn't the Chinese beat Europeans to the Scientific Revolution?—happens to be one of the few questions that people often ask publicly about why something didn't happen in history. It is analogous to the question of why your name did not appear on page 3 of today's newspaper. It belongs to an infinite set of questions that historians don't organize research programs around because they have no direct answers.

Sivin was a colleague of Needham and recognized the importance of making the largely overlooked contributions of Chinese science known to the world. Although Sivin downplays the question he certainly has invested a great deal of time and effort to highlight the achievements of Chinese science and dives deep into the cultural and historical aspects of Chinese science. Other sinologists, such as Justin Y Lin (Lin 1995), argue that the answer is in part related to the difference between what he calls 'Experience-based' and 'Experiment-based' trial and error.

> Experienced-based trial and error refers to spontaneous activity that a peasant, artisan, or tinker performs in the course of production. Experiment-based trial and error refers to deliberate, intense activity of an inventor for the purpose of inventing new technology. New technology obtained from experience is virtually free, while that obtained through experiment is costly.

Lin further argues that the gifted in ancient China had fewer incentives than their Western contemporaries to acquire the 'human capital' for modern scientific research. The 'human capital' he is alluding to is all of the time and effort that went into the civil service examinations. For many of the Chinese intelligentsia, the goal was to become a member of the ruling bureaucracy. To enter the civil service, scholars must distinguish themselves by passing the ordeal of the examinations. Although this system provided a way for people of 'low birth' to move up the socio-economic ladder, the ordeal was extremely time-consuming and left little opportunity for other intellectual pursuits. The exams required years of memorization of the Confucian Classics, and practice at writing poems and essays. Intense study of historical, philosophical and literary works occupied the thoughts of China's geniuses. Mathematical prowess and analytical skills were not subjects covered in the examinations. Thus, those intellectuals who had the mental acumen to conduct experimental investigations were busy with the examinations and the duties of bureaucratic life once they were successful.

When we ponder the Needham question, we should ask ourselves what Chinese 'scientists' thought they were doing. The term scientist is in quotes because the concept of scientist as we understand it was not known to them. They did not think of themselves as scientists and had no knowledge of that term as we would understand it. In our modern understanding of science and technology we often think of technology as a direct application of scientific research. Technology and invention flow from the results of scientific research. The view in ancient China is a very different one. Science in its early form was a matter for a small cadre of literati who were often not actually engaged in the hands-on technological process. Thus, technologies that were recorded in books were not written down by the practitioner, but rather, by a third party who commented on what was observed. As an experimental physicist, I perform the experiments and then write a paper detailing my work. Technology was something that the 'lower classes' engaged in. A technology was often a traditional way of doing something that was passed down, usually orally, through a family or to an apprentice. As Sivin points out (Sivin 1990):

> The sciences reflected the concerns of the tiny literate elite, their cosmologies, and the managerial problems they encountered in their careers and recorded in their writings. Technology was on the whole a matter of craft traditions, passed down privately from father to son or from master to apprentice. It was on these mainly oral and manual traditions rather than on cumulative science, recorded in writing, that the technological preeminence of China was built.

The purpose of bringing up the Needham question is not to give our readers an exact answer to the question. As we progress through the contributions of Chinese science, thoughtful readers may wonder and ask themselves the very same question. It is both comforting and instructive to know that you are asking the same question that has puzzled thinkers who came well before you.

Notes on the Chinese Language and Terms

Throughout this book we will use authentic Chinese terminology. As mentioned in the Introduction, we will use the simplified form of Chinese characters. This is the modern form of Chinese that is currently used in the People's Republic of China

(PRC). The PRC began character simplification in 1956. Adjacent to each Chinese character is a type of phonetic pronunciation guide known as Hanyu Pinyin Romanization. In modern China, it is not uncommon to see both character and pinyin used together. Chinese children learn both in school. Mandarin Chinese is the name that we give it in English. This term is not actually a Chinese word at all. The term is derived from the Portuguese term *mandarim*, which refers to ministers or court officials. Early Jesuit missionaries of the 16th century used this term to describe the court officials and the language that they spoke. Modern standard Chinese is based in the dialect spoken in Beijing. A Chinese person would more likely use the term 普通话 Pǔtōnghuà to refer to their language. Notice the pinyin word is written with tone marks and is actually a compound word formed from three other words. The meaning is more literally 'the common language'. Let's examine each of these individual words to learn a bit more of the Chinese language. The character 普 pǔ means universal, general, or widespread. The second character, 通 tōng can mean pass through, common, or communicate. Finally, 话 huà means speech, talk language, or dialect. In Chinese it is very common to use combinations of two characters to express an idea. The combination 普通 pǔtōng helps to narrow down the meaning to common, ordinary, average, or general. Language tends to be a unifying factor and the term 普通话 Pǔtōnghuà reflects the idea that this language unifies the country. China is a huge country and has many ethnic groups that speak their own languages or dialects. You will also notice the tone marks over vowels in each word. Chinese is a tonal language and the meaning of the word is determined by how the root sound is modified by intonation. A quick check in a Chinese dictionary will reveal more than thirty characters that have the root sound 'pu'. In the Beijing dialect there are four basic tones but sometimes you can encounter a fifth or neutral tone. Other dialects can have more than four tones. Cantonese, another variety of Chinese, is spoken mainly in South China. You will hear Cantonese in places like Guangzhou (Canton), Hong Kong and Macau. As if four tones were not enough, Cantonese has nine tones. The four basic tones have an intonation that resembles the tone mark above the vowel. Here are some examples with the root sound pu.

潽 pū (first tone) to boil over.

仆 pú (second tone) servant.

浦 pǔ (third tone) riverbank.

瀑 pù (fourth tone) waterfall.

As I mentioned there are some thirty others. The intonation is guided by the mark so pū has a flat tone. Pú is pronounced with a rise at the end like the way we ask a question in English. Pǔ goes down and then up. The fourth tone, pù, falls at the end like the tone we might use when being sharp or abrupt with someone.

Although this is not a Chinese language class, we do try to pronounce the terms in an authentic way. A keen observer, such as a scientist, might notice that 潽 pū, 浦 pǔ and 瀑 pù share some commonality in their shapes as well. Chinese characters are made up of subunits known as radicals. Over 80% of Chinese characters are phono-semantic compound words. That is to say, that there is a semantic part of the character that gives a broad meaning and a phonetic part that often gives a guide to the sound. Generally, one has to learn to pronounce each character from memory since characters do not make up a phonetic alphabet. There are only a little over 200 radicals that make up all of the simplified characters. A good dictionary will contain tens of thousands of individual characters.

Now look at 潽 pū, 浦 pŭ has a related sound but with a different tone. The right side of 瀑 pù does not look the same as the right side of 浦 pŭ, but it is another character that is pronounced as 暴 pù, bào, or bó. This character, 暴 pù, can be violent, brutal or tyrannical. Now look at the left side of all three characters. They have a common radical 氵. This is the radical for water, which is a slightly different form of the character 水 shuǐ. In fact 水, 氵 and 氺 all indicate water. All three words 潽 pū, 浦 pŭ riverbank and 瀑 pù waterfall have something to do with water.

仆 pú is also composed of two radicals. 亻, the radical on the left, is a compressed form of 人 rén, which means person. On the right we have 卜 bŭ, which means fortune telling or prophesy. The sounds B and P are similar consonants that have the same mouth position but differ in being voiced or unvoiced. P is an unvoiced consonant, meaning only air passes through the mouth while B is a voiced consonant. B requires the same mouth position but you have to vibrate your vocal cords to make the proper sound. B P sounds are often similar in many different languages. Again, this is not a language class, but it is interesting to see the underlying structure of Chinese characters. Studying the radicals that make up Chinese characters can sometimes provide insight into the meaning and story behind the character in much the same way as we study the Latin, Greek and Germanic roots of words in the English language. Chinese has many sounds that simply do not exist in English and are difficult for English speakers to reproduce.

Beginning Chinese language students are often frightened by the fact that words can have a very similar sound to the western ear, but rather different meanings. Even a native speaker may not recognize a word properly pronounced without context. Chinese contains many two word combinations. Fortunately, a native speaker hearing pŭtōng mispronounced is likely to understand the intended meaning. You can imagine the interesting word play possibilities that exist in such a language. There is a form of Chinese humor that uses this to its advantage.

A Mandarin Chinese Pinyin chart with audio can be found at the following web site: https://chinese.yabla.com/chinese-pinyin-chart.php. Example recordings demonstrate the pronunciation of each root sound with four different intonations.

Now let's apply our discussion to the title of the physics class we teach at Mercer University. We will not look into all the radicals but rather just learn how to pronounce the words. Several of the terms will show up again and again in our study of Chinese science.

Ancient Chinese Science and Technology-中国古代科学技术 Zhōng Guó Gŭ Dài Kē Xué Jī Shù

The first term 中国 Zhōng Guó is China. It does not sound like the word China in English and it literally means Middle Kingdom. This is a very old term used by the Chinese to describe their country. 中 Zhōng means middle or center and 国 Guó can mean country, nation or kingdom. This term has been found on a Bronze vessel dating back to the Western Zhou Dynasty (1046–771 BC). These characters are

Figure 1.1. Oldest known inscription of the characters 中国 Zhōng Guó. (Unknown, Western Zhou bronze inscriptions, Early Western Zhou Dynasty, via Wikimedia Commons; https://commons.wikimedia.org/wiki/File%3A%E4%BD%95%E5%B0%8A.jpg.)

shown in the highlighted box (figure 1.1). You will immediately notice that they do not look like modern simplified characters. They are a much older form of Chinese characters that tend to be a bit more pictorial.

古代 Gǔ Dài means ancient times. 古 Gǔ means old, classic or ancient and 代 Dài here means era or generation.

The combination 科学 Kē Xué means science and 学 means learning, knowledge or school.

技术 Jī Shù is a combination of two terms that mean technology. The individual characters that make up this phrase are 技 Jī, which means skill, ability, talent, ingenuity, and 术 Shù art, skill, special feat; method, technique. When we put it all together we get 'Ancient Chinese Science and Technology'.

Is it Really Science?

When we think about ancient Chinese science and scientists we must remember that the term scientist did not exist at the time period we are investigating. There was no such thing as the 'scientific method' and people were not trained as scientists. Much of what we call Chinese science can be found in books on philosophy or commentaries concerning observations on the natural world. There were books on mathematics, medicine, astronomy, mechanical design and other areas of what we would call science, but they are not organized or written in a style that we would expect as modern people. As I mentioned previously, sometimes the authors were only repeating knowledge that was handed down to them and they were often not the actual practitioner of the particular art or technique. For these reasons, some modern authors might dismiss the idea of ancient science, simply because it does not fit our modern standard. Our modern understanding of science often employs mathematics to describe physical phenomena. As modern scientists, we think of mathematical models and fundamental equations, such as $\mathbf{F}_{\text{net}} = m\mathbf{a}$, or $E = mc^2$. Certainly there were ancient books on mathematics, but many scientific ideas are conveyed as observations of the natural world. The theoretical underpinnings that were invoked by the ancients may seem like metaphysics or quackery to us. We need to keep this in perspective. When non-contact forces, such as electric and magnetic fields were first proposed, they were also thought of as something 'spooky' and metaphysical. People were accustomed to contact forces like a lever or a rope causing motion not some invisible agent like the electric field. Thus, I think it is instructive to go back and look at a few areas of early Western science before our modern intellectual snobbery gets the better of us.

It is common in our schools to explain the scientific method in such a way as to give the impression that one always starts with a hypothesis and then turns to experimentation in order to check that hypothesis. That is a quaint idea, but a careful study of the history of science might lead one to believe otherwise. Often, the first event that leads to a scientific theory is a simple observation or accidental discovery. Upon making such an observation, one then asks why something behaved in a particular way. Will it do it again? What if I change this factor? This is often the beginning of experimentation and the germ of a hypothesis. On the door of my laboratory there is a sign with the following quote from the 19th century scientist Joseph Henry: 'The seeds of great discoveries are constantly floating around us, but they only take root in minds well-prepared to receive them' (Philosophical Society of Washington 1877). When I enter the lab, I find it helpful to reflect on these words and I encourage my students to do the same.

Before venturing into the world of Chinese science let us examine a few thoughts and observations of men better known to modern students educated in the Western tradition. Benjamin Franklin was a keen observer of the natural world and a very early experimenter in the field of electricity. He is considered to have been a polymath and shares many characteristics with the ancient Chinese scientists we will encounter. 'Polymath' is an interesting word used to describe people who transcend any simple one-word description of their talents and interests. The term polymath or πολυμαθής literally means 'having learned much'. Such a person is an expert in many different fields. These may be very diverse fields such as science, mathematics, music and art. A more common term used today is 'Renaissance man'. Franklin is often thought of as one of the founding fathers of the United States and we think of him as a statesman and a diplomat. Yet, he was a printer, author, activist, humorist, inventor and scientist. Every school child can tell you about his foolish activities with a kite during a thunderstorm, but hardly any can tell you how he arrived at the idea that lightning and static electricity are related. His experiments contained no complex mathematical equations or data analysis. They were in fact simplistic and 'primitive' by today's standards. Franklin himself introduced the use of 'positive' and 'negative' to describe the electrostatic state of matter. Shown below is an extract from a letter by Franklin addressed to his fellow investigator Peter Collison (Franklin 1747).

Philada. May 25, 1747.

Sir.

In my last I informed you that In pursuing our Electrical Enquiries, we had observ'd some particular Phaenomena, which we lookt upon to be new, and of which I promised to give you some Account; tho' I apprehended they might possibly not be new to you, as so many Hands are daily employed in Electrical Experiments on your Side the Water, some or other of which would probably hit on the same Observations.

The first is the wonderful Effect of Points both in drawing off and throwing off the Electrical Fire. For Example,

Place an Iron Shot of three or four Inches Diameter on the Mouth of a clean dry Glass Bottle. By a fine silken Thread from the Ceiling, right over the Mouth of the Bottle, suspend a small Cork Ball, about the Bigness of a Marble: the Thread of such a Length, as that the Cork Ball may rest against the Side of the Shot. Electrify the Shot, and the Ball will be repelled to the Distance of 4 or 5 Inches, more or less according to the Quantity of Electricity. When in this State, if you present to the Shot the Point of a long, slender, sharp Bodkin at 6 or 8 Inches Distance, the Repellency is instantly destroy'd, and the Cork flies to it. A blunt Body must be brought within an Inch, and draw a Spark to produce the same Effect. To prove that the Electrical Fire is drawn off by the Point: if you take the Blade of the Bodkin out of the wooden Handle, and fix it in a Stick of Sealing Wax, and then present it at the Distance aforesaid no such Effect follows; but slide one Finger along the Wax till you touch the Blade, and the Ball flies to the Shot immediately. If you present the Point in the Dark, you will see, sometimes at a Foot Distance and more, a Light gather upon it like that of a Fire-Fly or Glow-Worm; the less sharp the Point, the nearer you must bring it to observe this Light: and at whatever Distance you see the Light, you may draw off the Electrical Fire, and destroy the Repellency. If a Cork Ball, so suspended, be repelled by the Tube, and a Point be presented quick to it, tho' at a considerable Distance, tis surprizing to see how suddenly it flies back to the Tube. Points of Wood do as well as those of metal, provided the Wood is not dry.

To shew that Points will throw off, as well as draw off the Electrical Fire: Lay a long sharp Needle upon the Shot, and you cannot electrise the Shot, so as to make it repel the Cork Ball. Fix a Needle to the End of a suspended Gun Barrel, so as to point beyond it like a little Bayonet, and while it remains there, the Gun Barrel cannot be electrised (by the Tube applied to the other End) so as to give a Spark; the Fire is continually running out silently at the Point. In the Dark you may see it make the same Appearance as it does in the Case before mentioned.

The Repellency between the Cork Ball and Shot is likewise destroy'd; 1. By sifting fine Sand on it; this does it gradually: 2. By breathing on it: 3. By

> making a Smoke about it from burning Wood: 4. By Candle Light, even tho' the Candle is at a Foot Distance: These do it suddenly. The Light of a bright Coal from a Wood Fire, and the Light of a red-hot Iron do it likewise; but not at so great a Distance. Smoke from dry Rosin dropt into a little hot Letter Founders Ladle under the Shot does not destroy the Repellency; but is attracted by both the Shot and the Cork-Ball, forming proportionable Atmospheres round them, making them look beautifully; somewhat like some of the Figures in Burnets or Whiston's Theory of the Earth.
>
> n.b. This Experiment should be made [in a closet] where the Air is very still.
>
> The Light of the Sun thrown strongly on both Shot and Cork by a Looking Glass for a long Time together does not impair the Repellency in the least. This Difference between Fire Light and Sun Light is another Thing that seems new and extraordinary to us.

Now let us examine these observations and see if we can extract some scientific ideas. We will interpret the results based on our modern understanding of electrostatic forces and fields. Franklin's work with 'points' gave him insights that led to the invention of the lightning rod.

First, we are told that the glass bottle must be clean and dry. The bottle is glass, which is a good insulator, and since it is clean and dry there are no conductive paths to bleed off the charge from the iron shot. Notice that the shot is iron, which means it is a good conductor. Any electrical charge we deposit on its surface will be free to move around and distribute itself on the outer surface of the shot. The cork ball has two important properties. Cork is not dense so the ball is lightweight and requires only a small force to deflect it. The other attribute of cork is that it is an insulator. When it is charged, the charge is not free to move around on its surface. The thread supporting the ball is made of silk, which is also a thin lightweight insulator. When the shot is electrified, both the shot and the cork have the same sign of charge. Since like charges repel, there will be a repulsive force between the cork and the shot. A recreation of this experiment is shown below in figure 1.2. Here the cork is replaced with a small piece of Balsa wood that has a conductive coating. The amount of charge determines the force between them as was worked out by Coulomb many years later. That is why the distance is 'more or less according to the Quantity of Electricity'. The bodkin is a sharp needle, like a knitting needle, that is mounted in a wooden handle. The wooden handle is also an insulator, but apparently not a perfect insulator as it allows some charge to flow through the body to the ground.

Charge is induced on the surface of the bodkin needle. Since the needle is metal, electrons will be able to move freely in it. If the shot has a positive charge, conduction electrons in the needle will be attracted to the end closest to the shot. If the shot had a negative charge then the electrons would be repelled in much the same way that the cork ball is repelled by the shot of like charge. We now know that the strength of the electric field produced by a curved metal surface is related to the radius of curvature or the sharpness of the curve. If two metal rods of different sharpness have the same charge, the electric field in the vicinity of the sharp point is much higher than the field the same distance from the blunt point. Franklin tells us

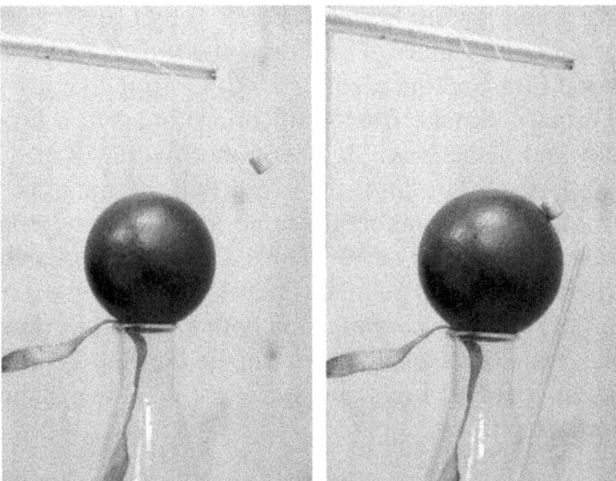

Figure 1.2. Balsa wood repelled by a charged metal ball (left). Repellancy is destroyed by the presence of a knitting needle (right).

that a blunt body must be brought closer to the shot to have the same effect. We are told that the sharp point will cause 'light gather upon it like that of a Fire-Fly or Glow-Worm'. In fact, the light that is produced is very much like lightning without the thunderous crack. We now know that air breaks down and becomes partially conducting once the electric field exceeds about 3×10^6 V m^{-1}. The result of this breakdown is the production of glowing plasma as shown below in figure 1.3.

Notice that we are told a blunt point must be brought closer to produce the light. The blunt point creates a smaller electric field near the tip compared to the sharp point at the same distance away. To create the light, the field must be at least as high as the breakdown strength of the surrounding air. The experiment with the sealing wax is very instructive. Sealing wax is an excellent insulator and charge is not able to bleed off to the ground. However, the moment the finger touches the metal needle there is a path to the ground. This in turn, drains away the charge and the repulsive force is decreased or even destroyed. When a sharp needle is placed upon the shot the same glow occurs. Ions are created by the intense electric field and electrons are transferred to the shot. This then neutralizes the charge on the shot destroying the repellency. Let's take a look at a few other ways in which the repellency is destroyed. The effect of smoke and candlelight is particularly interesting. Again we read:

> By making a Smoke about it from burning Wood: 4. By Candle Light, even tho' the Candle is at a Foot Distance: These do it suddenly. The Light of a bright Coal from a Wood Fire, and the Light of a red-hot Iron do it likewise; but not at so great a Distance.

What is common to all these cases in which something is burned? The answer is the production of ions. Of course, Franklin did not know about the production of ions; he was under the impression that the light produced from combustion was the

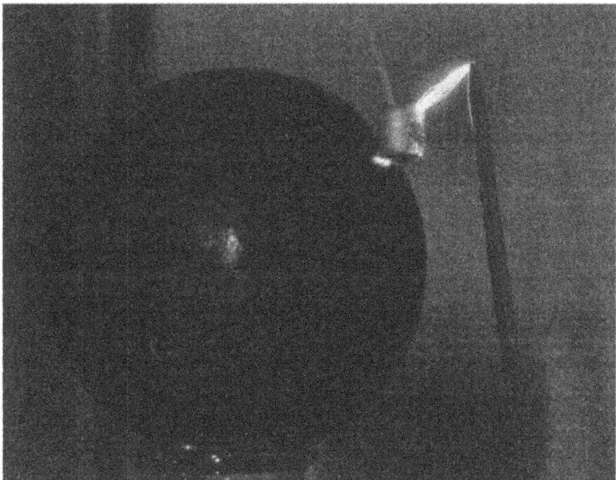

Figure 1.3. Discharge at the tip of a knitting needle.

cause. Notice that he refers to light in the description. A little later he makes it clear that he is thinking about light when he tells us:

> The Light of the Sun thrown strongly on both Shot and Cork by a Looking Glass for a long Time together does not impair the Repellency in the least. This Difference between Fire Light and Sun Light is another Thing that seems new and extraordinary to us.

It does seem a little strange that he did not realize that combustion products were produced in the case of the wood, coal and candle and no such products would have been associated with sunlight especially after being directed by a mirror. The nature of light was still not understood and he may have been thinking of light as some type of particle. Franklin was a great inventor, and in this experiment he is just a few steps away from inventing something that could have changed history and saved lives. He had a device that could sense combustion and that could have been turned into a fire alarm. Just think how a fire alarm system could have spared so much destruction of life and property. When the repellency was destroyed and the cork ball moved back, that motion could have been sensed by a mechanical device that could have triggered an alarm.

A second invention is also hiding in his description of the experiment.

> Smoke from dry Rosin dropt into a little hot Letter Founders Ladle under the Shot does not destroy the Repellency; but is attracted by both the Shot and the Cork-Ball, forming proportionable Atmospheres round them, making them look beautifully; somewhat like some of the Figures in Burnets or Whiston's Theory of the Earth.

Notice that in this case, the smoke particles are attracted to both the shot and the ball and the repellency is not destroyed. Ions are produced at high temperatures and here we only have smoke. Rosin itself is an insulator. Since they are attracted they either have the opposite charge or they might be acting as polarized dielectrics. In the case of a polarizable material, the charge centers are separated so that one side of the particle may have a positive charge and the other a negative charge. He does not actually say that they stick to the shot or the cork but form some type of atmosphere about them. The invention hiding in plain sight is the electrostatic precipitator. In Franklin's time, coal and wood smoke filled the air. Imagine the environmental impact and the respiratory problems associated with constant exposure to smoke. That invention had to wait until 1907 when Frederick Gardner Cottrell (Research Corporation for Science Advancement 2017) used an electrostatic device to collect sulfuric acid mist and lead oxide fumes. What if Franklin could have created a device that attracted smoke particles produced by all those chimneys in the same way he describes in the experiment? What he lacked at the time was a way to produce a continuous source of charge. In Franklin's era most devices that produced electric charge were based on friction. These friction machines, as they are known, required some sort of rotating mechanism that caused rubbing. They were usually hand-driven devices, not particularly suited as a source of continuous charge. The effectiveness of such machines was also limited by humidity and contamination of the frictional surfaces. It was not until 1800 when Alessandro Volta invented the voltaic pile, that a continuous non-mechanical source of electricity became available. We think of the voltaic pile as a battery, yet it was Franklin who introduced that term. As he used the term, it referred to a collection of Leyden jars used to store charge. They were grouped together like a battery of guns.

So why take this journey into Franklin's early work in electrostatics? It might seem out of place in a book on ancient Chinese science. Franklin is an example of someone who is not always looked upon as a scientist and who operated in a way that was rather different from the way modern students view the scientific method. Many of the early Chinese scientists were polymaths just like Franklin. Some were philosophers or served the emperor as bureaucrats. In many cases they reported observations of phenomena and gave explanations that might sound very unscientific to Western ears. Franklin did not rely on mathematical analysis. Apart from schematic diagrams, we do not see any graphs, curve fitting or derivation of equations as one might expect in modern scientific analysis. Yet what he did was science. Sometimes modern people look upon the discoveries of the ancients and dismiss their ideas as not being scientific. What is science? Scientists are trying to discover how the universe works. We look for unifying principles that describe the natural world. Sometimes we do this for the sole purpose of increasing man's understanding. Other times we might have a specific practical application or problem we are trying to solve. The way we conduct our research has been largely codified and our expectations for how we present our models of the universe have evolved. Yet, we still ask the same fundamental questions about how the universe works. The ancients might also ask why the universe works in a particular way. When you ask the 'why' question you move into an area of philosophy that makes many, but not all, modern scientists uncomfortable. Many would rather leave that to philosophers. This sharp dichotomy did not exist in ancient times.

Another theme we will notice in our study of Chinese science is how close they were to inventing something that we consider to be 'modern technology'. We may find ourselves saying 'If they just had such and such' or 'If they had just done this a little differently they would have developed the xyz'. That type of thinking brings us back to the Needham question.

Scientists make observations and try to understand how the universe works. Sometimes their ideas are expressed mathematically. Early scientists did not always have the mathematical tools to do so. Calculus and differential equations are relatively new inventions compared to the history of observation and attempts at understanding the natural world.

References

Bacon F 1960 *The New Organon and Related Writings (Library of Liberal Arts, no. 97)* ed F H Anderson (Englewood Cliffs, NJ: Prentice Hall)

Franklin B 1747 *Letter from Benjamin Franklin to Peter Collison* Retrieved from http://www.benjamin-franklin-history.org/letter-from-benjamin-franklin-to-peter-collison-dated-may-25-1747/

Lin J Y 1995 The Needham Puzzle: Why the Industrial Revolution did not originate in China *Econ. Dev. Cult. Change* **43** 269–92

Needham J 1969 *The Grand Titration: Science and Society in East and West* (Toronto: University of Toronto Press)

Philosophical Society of Washington 1877 Joseph Henry Presidential Address to the Philosophical Society of Washington 24 Nov. 1877 *Bulletin of the Philosophical Society of Washington* (Washington: Smithsonian Institution) p 163

Research Corporation for Science Advancement 2017 *About Frederick Cottrell* Retrieved from http://rescorp.org/about-rcsa/history/about-frederick-cottrell

Sivin N 1990 Science and medicine in Chinese history *Heritage of China: Contemporary Perspectives on Chinese Civilization* ed P S Ropp, p 169

Sivin N 1995 *Science in Ancient China: Researches and Reflections, Why the Scientific Revolution Did Not Take Place in China-or Didn't it?* (London: Taylor & Francis)

Williams S W 1871 *The Middle Kingdom; A Survey of the Geography, Government, Education, Social Life, Arts, Religion, etc., of the Chinese Empire and its Inhabitants* vol II, 4th edn (New York: Wiley)

Williams S W 1901 *The Middle Kingdom, A Survey of the Geography, Government, Literature, Social Life, Arts and History of the Chinese Empire and its Inhabitants* vol II (New York: Scribner)

Chapter 2

Kinematics (运动学 Yùndòngxué)

Now we will turn our attention to some early Chinese works that contain kernels of scientific ideas. We begin our journey with perhaps one of the most important texts in the history of Chinese logic and science known as the 墨子 (Mòzǐ). Students of Chinese philosophy are familiar with this text for its discussions of ethics, defensive fortifications and analogical argumentation. Yet it contains numerous short statements related to concepts of space, time, mechanics, geometry and optics. Much of the book is attributed to 墨子 (Mòzǐ), who is also known by the name 墨翟 (Mò Dí). We say 'attributed to' because it is not entirely clear how much of the work is by 墨翟 Mò Dí and which parts were written by his followers. In the West he is sometimes referred to by the Latinized version of his name, which is Micius. The dates of his life are not certain, but he probably lived c. 470–391 BC. This would make him a contemporary of Socrates. In ancient China this era was known as the Warring States period (战国时代 Zhànguó Shídài). Ancient Chinese uses very compact language and this makes translation very difficult. There are several English translations of the Mòzǐ and, not surprisingly, they are often very different. The translation problem is compounded by the fact that the text uses many characters with lost or uncertain meanings. Thus, copyists may have introduced new or erroneous meanings. This has led to numerous emendations. For all of its importance in the history of Chinese science it seems to be one of the most difficult works to accurately translate. Needham (Needham and Wang 1956) and his colleagues have translated some sections and they can be found in *Science and Civilisation in China*. Two other important translations are by A C Graham (Graham 1978) and Ian Johnston (Johnston 2010).

What is the most useful machine known to the ancient world? During the Renaissance scholars categorized various types of mechanical devices and found that there were six basic types of 'simple machines'. These are: the lever, wheel and axle, pulley, inclined plane, wedge, and screw. Most other mechanisms can be

doi:10.1088/2053-2571/ab03cbch2

thought of as a combination of their properties. All of these machines provide some form of mechanical advantage at the expense of a trade-off. A lever might require less force to lift a heavy load. The trade-off or expense is the distance of the lever arm compared to the distance that the load moves. The principle behind the lever is torque balance, which we will discuss in chapter 4.

Keep in mind that levers come in several different configurations. Most people think of using a long beam to lift a heavy weight like the lever shown below (figure 2.1). Here, the fulcrum rests on the ground and you apply some small force (*F*) to lift the load (*L*).

You can also hang a lever and make a weighing scale out of it (figure 2.2). Now the fulcrum is at the point of suspension and a small force (*F*) can lift a heavy load (*L*).

This type of arrangement is called a steelyard or Roman balance in the West and the Chinese call it a 秤 chèng.

In book 10 (卷十 juàn shí) of the 墨子 Mòzǐ we find a section that is sometimes translated as Canons (经 Jīng) and Explanations (经说 Jīng shuō). The Canon (经 Jīng) consists of short compact statements having to do with language, logic, and fragments of scientific thought. Each Canon is followed by an explanation, (经说 Jīng shuō); the purpose of which is to further explain or clarify the statement. There are two sections of Canons and Explanations 經上 (jīng shàng) and 經下(jīng xià), or Canon A and B as we will call them. Unless otherwise noted we follow the Johnston translation (Johnston 2010), which contains Canon **A1–A99** and **B1–B81**. The English text is given with the Canon reference number in bold face.

Concerning the lever and balance we find an interesting statement (**B25**), which we will explore.

> Bearing and not inclining. The explanation lies is 'being equal to'.
>
> Bearing: in the case of a horizontal piece of wood, if a weight is added to it and it does not incline, the counterweight 'is equal to' the (applied) weight. If the point of suspension of the counterweight is moved to the right and without adding to it, it does incline, the counterweight 'is not equal to' the weight. When horizontal, if a weight is added to one of its sides, it necessarily inclines

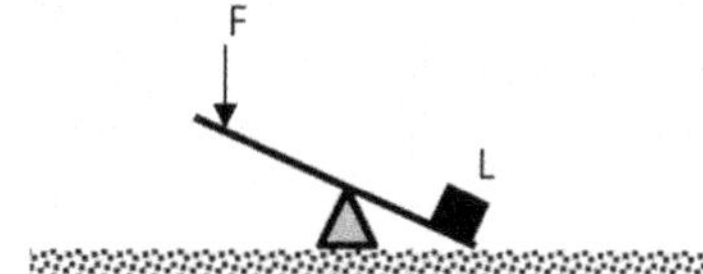

Figure 2.1. Conventional lever with fulcrum on ground.

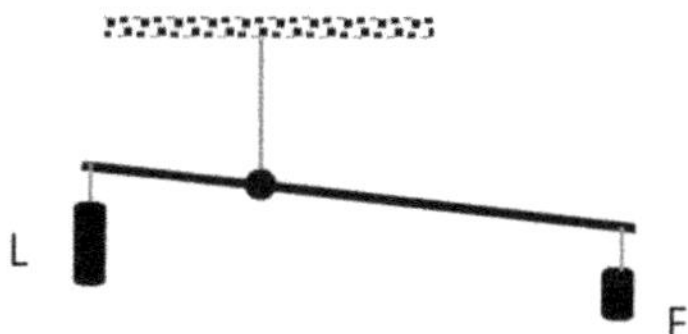

Figure 2.2. Lever with suspended fulcrum.

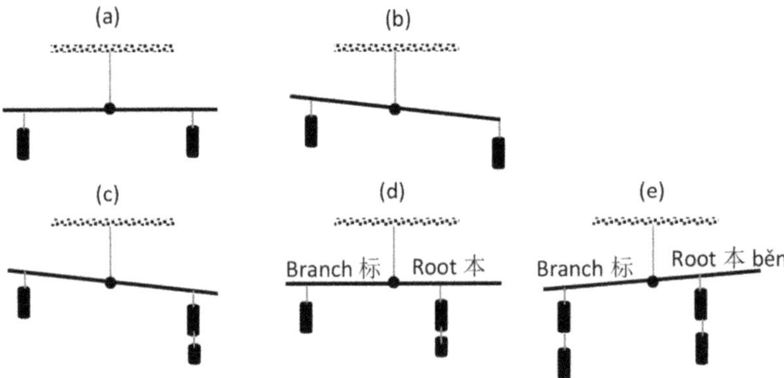

Figure 2.3. Lever with weights and distances modeling the description given in the 墨子 (Mòzǐ).

> downward, the counterweight and the weight corresponding (i.e. prior to the addition). When both are horizontal then the 'root' is short and the 'branch' is long. When the two are added to with equivalent weights, then the 'branch' necessarily falls, the 'branch' acquiring the 'force'.

The sequence of events described in the text is shown above (figure 2.3). In position (a) the wooden beam is horizontal and the two weights are the same distance from the point of suspension (fulcrum). Position (b) is different. Now, the point of suspension of the counterweight is moved to the right and without adding to it, it does incline. We have not changed the size of the weights only the position. The text tells us the counterweight is not equal to the applied weight. It is not the actual heaviness of the counterweight that is different but something has clearly changed. Position (c) is initially balanced like position (a) but an additional weight is added and the beam inclines. So the beam inclines under two conditions, one in which the weight is made bigger and one in which the position of the counterweight is changed without making it heavier. We can restore the balance as shown in position (d) if we move the point of suspension of the heavier side closer to the fulcrum. Thus, one side is short and the other is long. Position (e) shows us that if we keep one side short (root) and the other long (branch) and have equal weights on each side, the beam inclines downward on the long side. So it is clearly not a matter of just having equal weights on each side. Some other quantity is involved that depends on the position of the counterweights. We will learn about this quantity in chapter 4, which is about torque balance. While this is not a mathematical description of the phenomena it does give us an essential understanding of the way torque balance works.

Now let us examine a few other statements in the 墨子 Mòzǐ and see what we can learn about the motion (运动 yùndòng) of objects. Keep in mind that these statements are not intended to be a mathematical description of kinematics (运动学 yùndòngxué) but we use them to get students to think about how to describe motion. They give us some working definitions that we then modify and put into mathematical form. The translation is somewhat controversial and different translators render this differently. We are certain that it is a statement on motion but that is about as far as most translators will go. In Canon **A50** we find the following simple statement:

Movement is change of position.

Movement: Non-linear movement [rotation] is like a door hinge avoiding shutting.

Well, that seems obvious and we do not need a Chinese philosopher to tell us that when something moves it changes position. Let's look into that statement a little more deeply. If there is a change in position what else does that suggest? First, there must be a reference position to act as a basis for comparison. Second, we need a means of specifying what the position is. This implies that we must be able to make a measurement. In fact, all of what we do in physics requires the ability to make some type of measurement. Further, if you make a measurement, you need to invent measuring devices and standardize some type of measurement unit. In a more philosophical vein, Mohist philosophy was an analogical system. That is, it was based on analogies or comparisons between the nature of objects and ideas. Defining and compartmentalizing was an important part of the philosophical process. When you think about it, much of modern science is analogical. We compare theory to experiment and categorize different phenomena. Perhaps the most fundamental form of comparison is making a measurement. When we measure something, we compare the measured quantity to a standard unit.

In mathematical language we speak of the position of an object with respect to some origin or initial position. We can express the change in position as

$$\Delta x = x - x_0.$$

Here the x refers to a position along the x-axis and x_0 is our reference position. The change in position is also called the displacement (位移 wèiyí). Displacement is a tricky concept. If the displacement is zero it does not necessarily follow that the object has not moved. The object can move and return to its original starting position. In that case, the displacement is zero but the object has not remained motionless.

In this short discussion of displacement we have also assumed another quantity that we have not explicitly mentioned. That quantity is time. We say the object was 'here' and then it is 'there'. In describing the motion of an object we must also recognize time. Many philosophical systems have an underlying assumption about the way time flows and the idea that a cause must precede and effect. We find this same reasoning in the 墨子 Mòzǐ. The following are additional statements from the Mohist Canons. The corresponding Canon is in bold face.

A cause is that which obtains before something comes about. **A1**

In treating time, space and motion we find the following statements:

A beginning is a specific instant of time.

A beginning: time in some cases has duration and in some cases does not. A beginning is a specific instant of time without duration. **A44**

Space involves movement in location. The explanation lies in the length in terms of both extension and duration (space and time). **B13**

Traveling a distance involves duration. The explanation lies in before and after.

Traveling: the one traveling is necessarily near before and far after. Far and near are distances. Before and after are durations. If people travel distances, it must involve duration. **B63**

If an object travels a distance it does not do so instantaneously. We can quantify the relationship between the distance traveled and the time required to travel that distance as velocity or speed. Velocity (速度 sùdù) and speed (速率 sùlǜ) are, however, not the same thing. Velocity concerns the displacement of the object and speed the total distance traveled. If an object moves 1.0 m in a straight line from a reference position and then returns to its starting point, the total distance traveled is 2.0 m. The displacement, however, is zero. This may seem strange at first but in order to get back to where it started from the object must have moved in one direction and then in exactly the opposite direction. In this way the velocity averages out to zero. Mathematically we express this relationship as follows:

$$\text{Speed} = \frac{\text{distance}}{\text{time interval}}.$$

And as for velocity we have,

$$\text{Average velocity} = \frac{\text{displacement}}{\text{time interval}}.$$

The average velocity (平均速度 píngjūnsùdù) is written more mathematically as

$$V_{\text{avg}} = \frac{\Delta x}{\Delta t},$$

where the Δ means the change in some parameter. We interpret Δx as the change in the position during the time interval Δt. That is $\Delta x = x_2 - x_1$ and $\Delta t = t_2 - t_1$.

Before we go further into this discussion we should mention the measurement of time. Many books have been written of ancient methods of measuring time. One time measuring device found worldwide in many different ancient cultures is the clepsydra or water thief. This device measures time based on the flow of water into or out of a time-indicating container. In China these devices were called 滴漏 dīlòu or 漏壶 lòuhú. Both words contain the character 漏 lòu which means to leak or drip. The main problem with such a water clock is that the rate of flow is not constant. As water is drained from the supply vessel, the pressure on the spout that leaks the water changes. Thus, the flow of water is not uniform in time. The device shown on the printed cover of this book is a modern day version that employs a floating siphon to correct for this effect (figure 2.4).

While visiting the NRI in 2012, we had an opportunity to experiment with this device. The floating siphon does seem to linearize the flow rate. Previously unpublished data is shown below in figure 2.5.

For about the first 100 s, both curves are linear but the flow rates are clearly different. The siphon mechanism provides an amazingly linear flow. The floating siphon is not found in any ancient Chinese versions of water clocks and this is a

Figure 2.4. A modern experimental water clock that employs a floating siphon to correct for pressure loss.

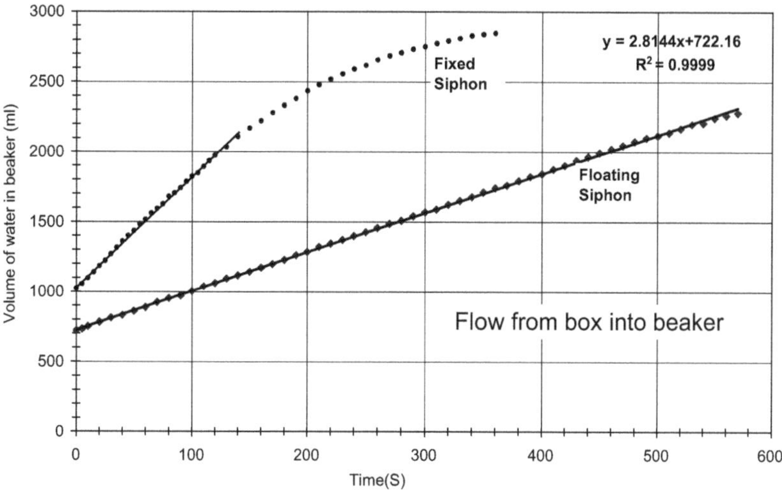

Figure 2.5. Volume of water collected in a beaker as a function of time. The straight line is for the floating siphon and the curved line is for a fixed siphon connecting the box to the beaker.

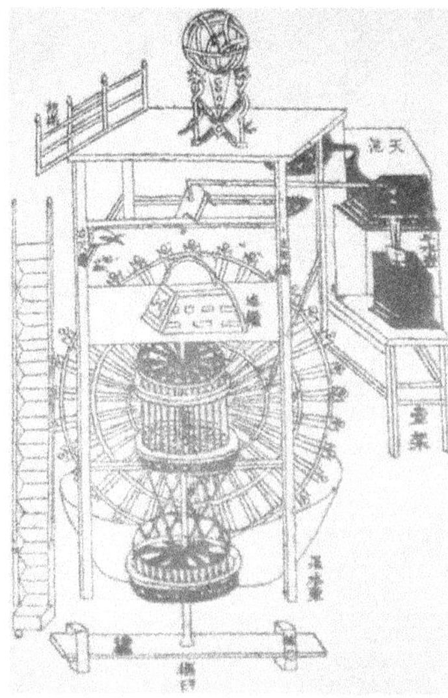

Figure 2.6. Su Song's astronomical clock tower described in 1092. (Su Song's Clock Tower from Song's Book, 2007, via Wikimedia Commons.)

modern modification. One of the most complex versions of a water clock is shown in figure 2.6. This is the great astronomical clock tower built by Su Song (苏颂 Sū Sòng), the great polymath of the Song dynasty (宋朝 Sòng cháo). The design of this clock was officially printed in 1094.

In the upper right corner of this figure we can see water flowing. The great wheel was driven by the weight of water captured by containers on the outer rim. On top of the tower is an armillary sphere that indicated the position of celestial bodies. The tower functions as a clock, calendar and astronomical indicator. Quite an amazing device for over a thousand years ago.

Sometimes we can analyze the motion of an object graphically. We can make a graph of the displacement of an object as a function of time. The graph shown in figure 2.7 illustrates this. Suppose we wish to find the average velocity of the object between 0.200 s and 0.400 s. From the graph know that at time $t_1 = 0.200$ s the object is at $x_1 = 0.0650$ m, and at $t_2 = 0.400$ s the object is at $x_2 = 0.210$ m.

We can then express the average velocity as

$$V_{\text{avg}} = \frac{\Delta x}{\Delta t} = \frac{x_2 - x_1}{t_2 - t_1} = \frac{0.210\text{ m} - 0.0650\text{ m}}{0.400\text{ s} - 0.200\text{ s}} = 7.25 \times 10^{-1}\text{ m s}^{-1}.$$

There are a few things to notice about this answer. First, we always check the units. As expected, the units come out in meters per second (m s^{-1}). We must also check the number of significant figures in the answer. Each number is expressed using three significant figures. The distance 0.0650 m has only three significant figures. The first zero after the decimal point is only a place holder and is not counted as a significant digit. Generally, leading zeroes are not significant. As a place holder it just tells us that the number starts in the hundredths place. Trailing

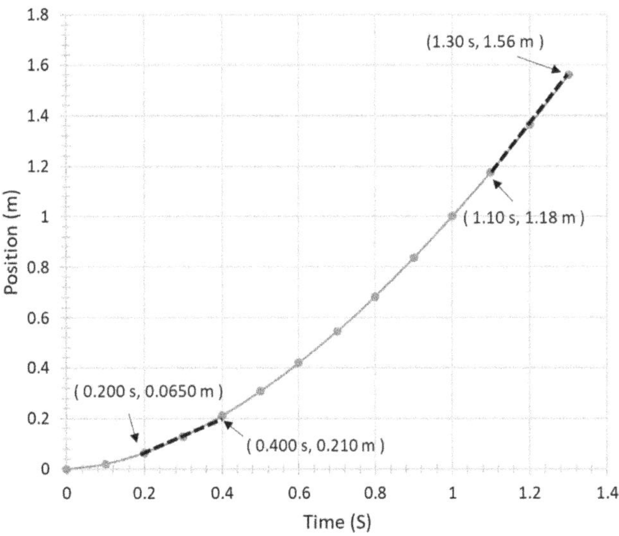

Figure 2.7. Displacement as a function of time.

zeros are significant and so is the zero after the five. That last zero tells us that we know with certainty that there is nothing in the ten thousandths place. We can simplify questions about significant figures by using scientific notation. Leading and trailing zeroes are eliminated when numbers are written in proper scientific notation. Thus, 0.0650 m should be written as 6.50×10^{-2} m, which clearly has three significant figures. Notice that 7.25×10^{-1} m also has three significant figures. The number of significant figures in the answer cannot exceed the number of significant figures in the data that went into the calculation. If there were more significant figures in the answer then the simple act of calculation would have created precision that was not originally in the data.

Anyone familiar with graphs will also notice that the slope of the position vs time graph is the same as the average velocity. The most common definition of slope (斜率 xiélǜ) is

$$\text{slope} = \frac{\text{rise}}{\text{run}}.$$

Rise is the change on the vertical axis, which is $\Delta x = x_2 - x_1$, and run is the change along the horizontal axis, which is $\Delta t = t_2 - t_1$. This fits the general form of the slope. If the slope of the displacement vs. time curve is increasing with time, then the velocity is increasing, and if it is decreasing with time then the object is slowing down. What about the case of constant velocity? That means that the velocity does not change with time. In that case the slope of the curve is constant. If the slope is zero then the velocity is zero and the object maintains its position. This results in a curve that is just a flat line. For the object shown in figure 2.7 the slope of the curve is changing with time. In the interval from 1.10 s to 1.30 s the slope of the curve is steeper than it was during the interval from 0.200 s to 0.400 s. This means that the

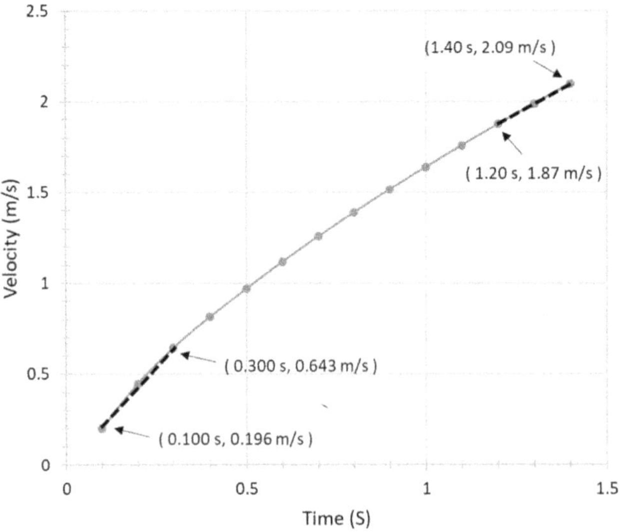

Figure 2.8. Velocity as a function of time.

object is moving faster during the later time interval. From the data on the graph we find that the average velocity during the later time interval is 1.90 m s^{-1}.

We can also draw a graph of the velocity of the object as a function of time. Such a graph is shown in figure 2.8.

What can we learn about the motion of the object from this graph? Just as the slope of the displacement vs. time curve gave the velocity of the object, we can examine the slope of the velocity vs time curve. In this case the slope yields the acceleration (加速度 jiāsùdù) of the object. The average acceleration during the first time period is given by

$$a_{\text{avg}} = \frac{\Delta v}{\Delta t} = \frac{v_2 - v_1}{t_2 - t_1} = \frac{0.643\text{ m} - 0.196\text{ m}}{0.300\text{ s} - 0.100\text{ s}} = 2.24\text{ m s}^{-2}.$$

It is clear from the curve that the average acceleration of the object is not the same during the interval from 1.20 s to 1.40 s. The slope of the curve is not quite as steep. During this time interval the acceleration is only 1.10 m s^{-2}.

Not only does the slope of the curve provide information but we can also learn something from the area under the curve. Figure 2.9 shows a magnified portion of figure 2.8.

The area under the velocity vs time curve gives the displacement of the particle. We can approximate the area under the curve by a triangle sitting on top of a rectangle as shown in figure 2.9. Area 1 represents the area of a triangle and area 2 is the area of a rectangle.

The area of a triangle is given by the equation

$$A = \frac{1}{2}B \times H,$$

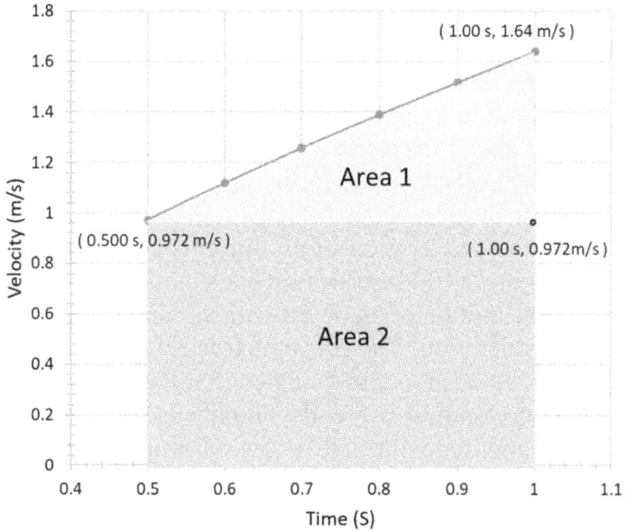

Figure 2.9. The area under a velocity vs time graph gives the displacement.

where B is the length of the base and H is the height of the triangle. Using the data points on the graph we see that the base is

$$B = 1.00\ \text{s} - 0.500\ \text{s} = 0.500\ \text{s},$$

and the height is

$$H = 1.64\ \text{m s}^{-1} - 0.972\ \text{m s}^{-1} = 0.668\ \text{m s}^{-1}.$$

The area of the triangle section is then

$$A_1 = \frac{1}{2} B \times H = 1.67 \times 10^{-1}\ \text{m}.$$

Keep in mind that although we are calculating an area, it should not have units of m^2 such as you would obtain if each dimension were in units of m. Notice the units are $(\frac{\text{m}}{\text{s}}) \times (\text{s}) = \text{m}$, which is what we expect for a displacement.

To obtain the area under the curve we have to include the area of the rectangle. For the rectangle the area is simply

$$A_2 = 0.500\ \text{s} \times 0.972\ \text{m s}^{-1} = 4.86 \times 10^{-1}\ \text{m}.$$

The total area is $A_1 + A_2 = 6.53 \times 10^{-1}\ \text{m}$.

This total area is the displacement of the object from time 0.500 s to 1.00 s.

There is a slight gap between the line on the graph and the edge of the triangle. The velocity vs time curve is not a straight line and the slope is clearly changing. Any straight segment laid along the curving line will not exactly match the shape and there will be a gap. If we make the straight segment progressively shorter and shorter, the gap will decrease. This means that the short segments are a better approximation to the line. In the language of calculus we are taking a limit as the time interval approaches zero. In finding the slope we can make the line segment shorter, by decreasing the time difference over which we do the calculation. The slope we obtain with an infinitesimally short time gives us the slope of a line tangent to the curve. The acceleration we obtain from this method is called the instantaneous acceleration. We can evaluate the slope of the displacement vs time curve using the same method of progressively shortening the time interval. The resulting slope in that case would give the instantaneous velocity.

In many situations the acceleration of an object is constant or nearly constant. The graph shown in figure 2.10 illustrates such a situation.

Figure 2.10 shows the motion of an object with varying acceleration compared to the case of constant acceleration. The constant acceleration case corresponds to the slope of the velocity vs time curve also being constant. Analyzing the motion of an object with constant acceleration is greatly simplified. We can relate the position, velocity, acceleration and time through a set of simple equations as shown in table 2.1.

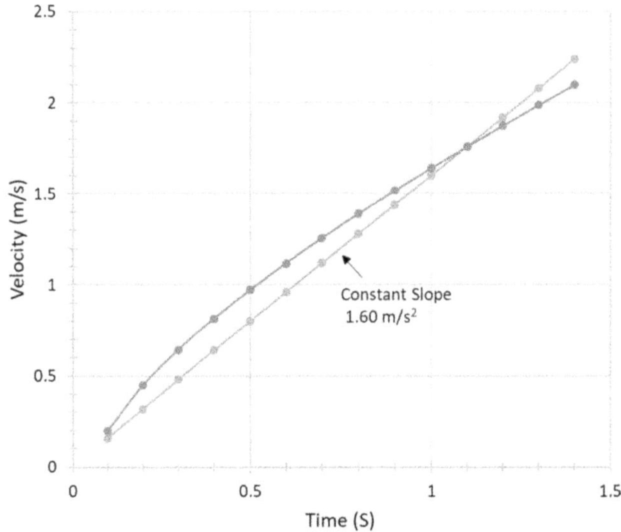

Figure 2.10. Velocity vs time with constant acceleration (slope).

Table 2.1. Equations for constant acceleration.

(2.1)	$v = v_0 + at$
(2.2)	$x = x_0 + v_0t + \frac{1}{2}at^2$
(2.3)	$v^2 = v_0{}^2 + 2a(x - x_0)$
(2.4)	$x = x_0 + \frac{1}{2}(v_0 + v)t$
(2.5)	$x = x_0 + vt - \frac{1}{2}at^2$

When encountering new equations it is always a good idea to interpret them. That is, to put them in words and explain the meaning of each term in the equation. All too often students just try to memorize equations without actually understanding what they mean. You should try to understand the function of each term in the equation and under what conditions it applies. Equation (2.1),

$$v = v_0 + at, \tag{2.1}$$

is nothing more than the very definition of acceleration just solved algebraically for the velocity. This equation tells us that if we know the acceleration and the initial velocity, we can then predict the velocity at some later time t. The initial velocity v_0 is the velocity the object has at the instant we start measuring the time ($t = 0$ s). Since the object is accelerating, the velocity will have changed and have a new value at some later time. If the initial velocity v_0 is positive and the acceleration is positive, then at a later time t, the new velocity will be greater than the initial velocity. If the acceleration is negative then the at term subtracts from v_0 and the velocity is less than v_0 at a later time. Sometimes an object will start from 'rest' meaning that it is not initially moving. In that case we set $v_0 = 0$ m s^{-1}.

The next equation in the table is (2.2), which enables us to find the position of the object at some time t after it has started moving. In the equation

$$x = x_0 + v_0 t + \frac{1}{2}at^2 \tag{2.2}$$

x_0 represents the initial position of the object at $t = 0$ s and v_0 gives the initial velocity. As time progresses the object changes its position from x_0 to a new location x. Since the object is accelerating, this new location is not just the initial velocity multiplied by the time ($v_0 t$). If the acceleration and the initial velocity are both positive, then as time progresses the object is speeding up. That means in equal intervals of time it covers a greater distance. The $\frac{1}{2}at^2$ term accounts for this extra velocity. As the time increases this term grows very quickly. In fact, it grows as the square of the time. That means if we double the time interval $\frac{1}{2}at^2$ it is not simply twice as big but is four times larger since $2^2 = 4$. If we triple the size of the time interval, the $\frac{1}{2}at^2$ term is now $3^2 = 9$ times larger.

The graph shown in figure 2.11 illustrates how the position changes as a function of time. In this case the initial position is $x_0 = 1.00$ m, $v_0 = 2.00$ m s^{-1} and $a = 4.00$ m s^{-2}. When we start timing the motion of the object it is initially at $x_0 = 1.00$ m and is already moving with a velocity of $v_0 = 2.00$ m s^{-1}. In the first 0.50 s it moves from $x_0 = 1.00$ m to $x = 2.50$ m for a change of 1.50 m. In the last 0.50 s ($t = 1.50$ s to $t = 2.00$ s) it travels from $x = 8.50$ m to $x = 13.0$ m. Now it covers a distance of 4.50 m in the same amount of time.

In this discussion we have been looking at the object accelerating in time. We can also determine how its speed changes as it travels over some distance. Relating the velocity to the distance traveled while accelerating is the job of the equation

$$v^2 = v_0{}^2 + 2a(x - x_0). \tag{2.3}$$

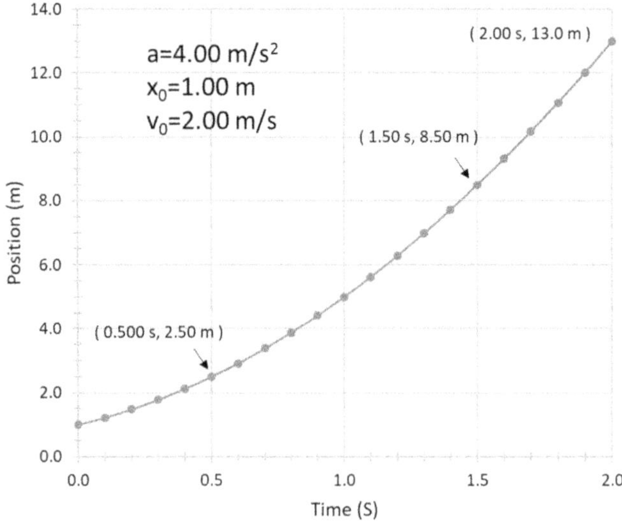

Figure 2.11. Position as a function of time with constant acceleration.

If we know how far the object traveled $(x{-}x_0)$, while being acted on by the acceleration a, we can then calculate the velocity after it covers that distance. A good example of this motion might be accelerating in a car. You are traveling along the highway with an initial velocity v_0 and you see a traffic jam ahead. There is a truck stopped a short distance in front of you. As you are slowing down to a stop you are not thinking so much about the time it takes. You are thinking about the distance between you and the stationary truck. That distance is getting shorter. Will you manage to stop before hitting the truck? In this case the acceleration a is a negative quantity. If you come to a stop your velocity at the position x is $v = 0$. Initially you were moving at a velocity v_0 when you were at position x_0.

The two remaining equations also allow us to find the position of an object but the variables are arranged a little differently. In equation (2.4) we are missing the acceleration term. The equation tells us how to find the position using the average velocity the object has during some time interval. The average velocity shows up as $\frac{1}{2}(v_0 + v)$ in the full equation

$$x = x_0 + \frac{1}{2}(v_0 + v)t. \tag{2.4}$$

Notice that the acceleration term is missing and we only use the time it has been accelerating and the average velocity the object has during that time interval.

The last equation (2.5) looks very similar to equation (2.2) but there is a difference.

$$x = x_0 + vt - \frac{1}{2}at^2 \tag{2.5}$$

does not contain the initial velocity v_0. Instead it has the velocity of the object at the instant of time t. If the object is accelerating, you cannot just find the distance traveled by using $x = x_0 + vt$. That would over count the total distance since v is not constant. In order to correct for that, we must subtract the additional displacement using the $-\frac{1}{2}at^2$ term.

With these equations we can solve many types of kinematics problems that have a constant or nearly constant acceleration.

References

Graham A 1978 *Later Mohist Logic, Ethics and Science* (Hong Kong: The Chinese University Press)

Johnston I 2010 *The Mozi, A Complete Translation* ed I Johnston (Transl.) (New York: Columbia University Press)

Needham J and Wang L 1956 *Science and Civilisation in China, History of Scientific Thought* vol 2 (Cambridge: Cambridge University Press)

IOP Concise Physics

Teaching Physics through Ancient Chinese Science and Technology

中国古代科学技术与物理教学

Matt Marone

Chapter 3

Force (力 lì)

In this chapter we will briefly discuss the concept of force (力 lì). In the previous chapter on kinematics we examined a few statements from the 墨子 Mòzǐ. We open this chapter with another statement, which at first seems simple but has a long history of misunderstanding. Following Johnston's translation (Johnston 2010) we read in Canon **A21**:

Force is what moves a body.

Seems simple, doesn't it? A body moves because a force acts on it. What about a situation where there does not seem to be a force propelling the object forward? Why does a body continue in motion after the force that started the motion is no longer in contact with the object? For example, you fire an arrow from a bow. The bow string exerts a force on the arrow and sends it on its way. What happens when the arrow is no longer in contact with the bow? In that case there is no obvious contact force so why does the arrow continue to move? Perhaps we are examining this statement too closely. It is not intended to be a law of motion, but it does raise some questions. The Mòzǐ is not the only ancient document that gets a bit muddled when it comes to this concept. Shortly after Mo Di's (墨翟 Mò Dí) time, and in a totally different culture, we find Aristotle struggling with a similar concept. In Aristotle's *Physics, Book VII*, we find the following statement (Aristotle 350 BC):

> Everything that is in motion must be moved by something. For if it has not the source of its motion in itself it is evident that it is moved by something other than itself, for there must be something else that moves it.

So what is keeping that arrow moving? It must be moved by something other than itself, but what? There were some quaint ideas about how the low pressure created behind the arrow caused air to rush in and kick the arrow, but something is still missing in the explanation.

Now we jump forward in time to 1687 and how Newton (牛顿 Niúdùn) expressed his understanding of the situation (Newton 1846).

> Law I.
>
> Every body perseveres in its state of rest, or of uniform motion in a right line, unless it is compelled to change that state by forces impressed thereon.
>
> PROJECTILES persevere in their motions, so far as they are not retarded by the resistance of the air, or impelled downwards by the force of gravity. A top, whose parts by their cohesion are perpetually drawn aside from rectilinear motions, does not cease its rotation, otherwise than as it is retarded by the air. The greater bodies of the planets and comets, meeting with less resistance in more free spaces, preserve their motions both progressive and circular for a much longer time.

This is the famous first law of motion which tells us that it is natural for the motion to continue and it only changes when some force acts to cause a change. Today we call this concept inertia (惯性 guànxìng). It is also worth looking at the second and third laws and how they were expressed simply in words. Newton did give mathematical equations to back up these laws but the form is different than that to which a modern reader is familiar with. His arguments were more along the lines of the geometrical arguments of Euclid. The second and third laws are as follows:

> Law II.
>
> The alteration of motion is ever proportional to the motive force impressed; and is made in the direction of the right line in which that force is impressed.

> Law III.
>
> To every action there is always opposed an equal reaction: or the mutual actions of two bodies upon each other are always equal, directed to contrary parts.

This discussion shows how our understanding of the natural world has evolved in time. Even though the 墨子 Mòzǐ does not give a precise mathematical statement, we should not underestimate the wealth of knowledge this work holds. 墨家 Mòjiā, or the school of Moism, produced some of the finest scholars of the time. They were experts in philosophy, logic and science. In later years they were seen as philosophical rivals to China's other great philosophical schools of thought including Confucianism (孔教 Kǒngjiào). Mohist teachings emphasize defensive warfare and the 墨子 Mòzǐ has a very long section on how to defend cities against various forms of attack. Mohist engineers were highly sought after as experts in defense, particular when the forerunners of the Qin dynasty (秦朝 Qíncháo) were conquering smaller states in an effort to unify China.

Table 3.1 lists a few physics concepts discussed in the 墨子 Mòzǐ. Again, they are not mathematical, but rather observations and descriptions.

Table 3.1. Physics-related concepts discussed in the 墨子 Mòzǐ.

Physics concepts	Canons and explanations A, B (经上,经上下)
Atwood's machine, lifting and lowering	B26
Lever and steelyard balance	B25
Force	A21
Movement	A50
Convex mirrors	B24
Image formation from mirrors	B22
Light and shadows	B19, B20, B21
Multiple light sources	B17, B18
Space movement and length	B13, B14
Local noon	A57
Time	A44
Stopping	A51
Cause and effect	A1
Traveling-velocity	B63

Now it is time to move our discussion of forces to a more mathematical formalism. We can express the second law as

$$\mathbf{F}_{net} = m\mathbf{a}.$$

Both **F** and **a** are vectors, which is why they are written in bold face. Vectors (矢量 shǐliàng) are sometimes expressed with an arrow over the top $\vec{F}$. A vector has both a magnitude and a direction. When we express just the magnitude, we simply write the vector without bold face or an arrow as in $F_{net} = ma$.

Mass (质量 zhì liàng) is not a vector but rather a scalar (标量 biāoliàng). It only has magnitude, but no direction. Notice in the vector the net force (净力 jìng lì) and acceleration (加速度 jiāsùdù) are in the same direction. The scalar mass can only change the magnitude but not the direction. Mass is always a positive quantity. As Newton said 'The alteration of motion is ever proportional to the motive force' and we can see in the equation that F is proportional to a. The constant of proportionality that relates F to a, is in fact the mass. The fact that F and a are in the same direction is consistent with the statement 'and is made in the direction of the right line in which that force is impressed'. One way to think about mass is that it relates the acceleration you obtain, to the force you apply to an object or,

$$m = \frac{F_{net}}{a}.$$

The units of force are named after Newton. Since force is mass multiplied by acceleration we have $1\ N = 1\ kg\ m\ s^{-2}$.

We also must consider the net force acting on the object. The net force is the vector sum of the forces acting on the object. Two forces can act on the object but be equal in magnitude and opposite in direction. In that case the net force would be zero. If the net force acting on an object is zero, then this implies that the acceleration a is also zero.

$$\mathbf{F}_{\text{net}} = 0, \; \mathbf{a} = 0.$$

When $\mathbf{a} = 0$ this does not mean that the object is not moving. It does mean that it is not accelerating or changing its motion. If it is moving with velocity v it will continue to do so and if it is stationary ($v = 0$), it will remain that way. This carries the idea of Newton's first law and inertia.

The way we add force vectors to find the net force is geometrical. A vector is made up of parts called components. The components of the vector are aligned along the axes of a coordinate system as shown in figure 3.1.

The vector $\mathbf{F}$ has a projection along the x-axis that is the component F_x and a projection along the y-axis that is F_y. These components are related to the magnitude of the vector $\mathbf{F}$ through the trigonometric identities $\sin\theta$, $\cos\theta$, $\tan\theta$. From the geometry of the vector we have

$$\cos\theta = \frac{\text{adjacent}}{\text{hypotenuse}} = \frac{\mathrm{F}_x}{\mathrm{F}}, \quad \sin\theta = \frac{\text{opposite}}{\text{hypotenuse}} = \frac{\mathrm{F}_y}{\mathrm{F}}.$$

We can then solve for the x- and y-components of the vector $\mathbf{F}$,

$$\mathrm{F}_x = \mathrm{F}\cos\theta \text{ and } \mathrm{F}_y = \mathrm{F}\sin\theta.$$

If you know the components of the vector, then you can find the magnitude from the Pythagorean theorem

$$\mathrm{F} = \sqrt{\mathrm{F}_x^2 + \mathrm{F}_y^2}.$$

If more than one force acts on the object we must find the net force in each direction. We do this by adding the corresponding components together so that

$$\mathrm{F}_{\text{net}x} = F_{1x} + F_{2x} = ma_x, \; \mathrm{F}_{\text{net}y} = F_{1y} + F_{2y} = ma_y.$$

This gives us a way to calculate the acceleration along both the x- and y-axes. Let's analyze the example shown below. We have an object of mass 10.0 kg that moves on a horizontal low friction surface. Two forces act on the object. Force F_1 is 5.00 N at an angle of 30.0° above the horizontal. Force F_2 is 2.00 N and acts

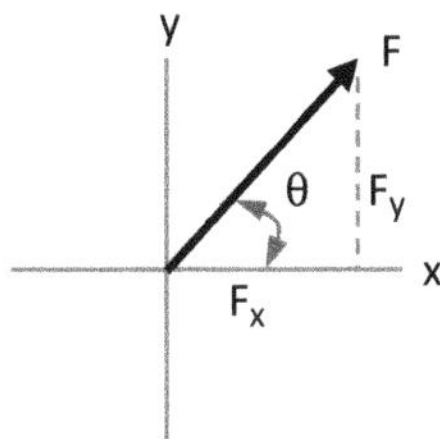

Figure 3.1. Components of a vector.

horizontally but in the negative x-direction. This is shown below in figure 3.2. Notice on the left side of this figure that we have a physical picture of the situation. On the right-hand side we have what is known as a free-body diagram. The object has been replaced by a single dot and all the forces act on that dot. Also notice there is a vector showing the acceleration. This is not a force, so it does not touch the dot, it only serves as a reference direction. The free-body diagram also includes a force N and the weight W (重量 zhòngliàng). The force N is called the normal force (正向力 zhèngxiànglì). Remember Newton's third law tells us that that the weight pushing down is opposed by the surface that supports the object. Sometimes students have the mistaken idea that this force is not real or unnecessary. If the object was not acted on by the other forces of 5.00 N and 2.00 N, only N and W would be acting. The object is not accelerating in the y-direction. If we ignored the normal force and only counted the weight, how could we have a net force of zero? If we include the normal force then in the y-direction we would have

$$F_{nety} = N - W = 0, \text{ not the impossible situation of } F_{nety} = -W = 0.$$

When dealing with multiple forces, we use a force table to help organize our work. Table 3.2 shows all the acting forces and their components. Two columns have the x- and y-components in equation form. Once the components are represented, then we just sum all the terms in each column. Writing everything in equation form helps us to find errors and is more efficient than carrying a lot of numbers line after line in a long calculation.

The forces N and W only act in the y-direction so their x-components are zero. At this point the force N is unknown so we treat it as a variable. The weight W acts in the negative y-direction so we insert the negative sign. The force F_2 acts in the negative x-direction and has no y-components. For the weight we used $W = mg$ which acts in the negative y-direction.

The net force is found by summing all the terms in the corresponding column. This yields

$$F_{netx} = F_1 \cos\theta - F_2 = ma_x \text{ and } F_{nety} = F_1 \sin\theta + N - mg = 0.$$

We set $F_{nety} = 0$ because we were told that the object only accelerates in the horizontal direction.

Now we can solve for the horizontal acceleration.

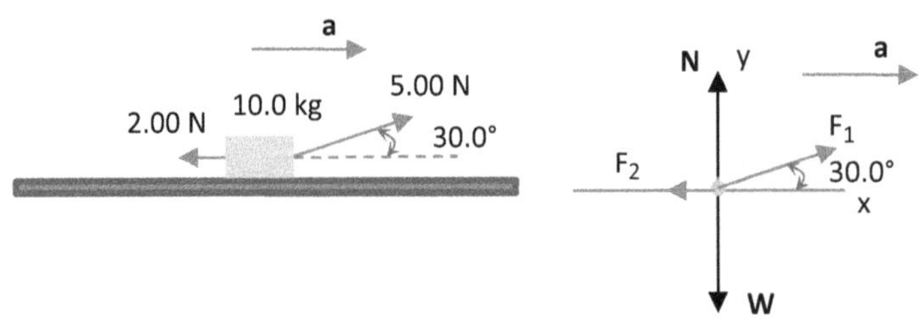

Figure 3.2. Object moving under the influence of several forces. Physical picture (left) and free-body diagram (right).

Table 3.2. Force components.

Force	x-component	y-component
F_1	$F_1 \cos\theta$	$F_1 \sin\theta$
F_2	$-F_2$	0
N	0	N
W	0	$-mg$

$$\frac{F_1 \cos\theta - F_2}{m} = a_x = \frac{5.00\ \mathrm{N} \cos 30.0^\circ - 2.00\ \mathrm{N}}{10.0\ \mathrm{kg}} = 2.33 \times 10^{-1}\ \mathrm{m\ s^{-2}}.$$

From the y-components we have

$$N = mg - F_1 \sin\theta = 10.0\ \mathrm{kg} \times 9.80\ \mathrm{m\ s^{-2}} - 5.00\ \mathrm{N} \sin 30.0^\circ = 9.55 \times 10^1\ \mathrm{N}.$$

Notice that the normal force is not simply the weight, but it is a little less, since the force F_1 has a component that slightly supports the object.

In the preceding problem we had a 2.00 N force acting in the negative x-direction. Although we did not explicitly state the source of this force, it could very well have been caused by friction (摩擦力 mócālì). Friction always points in a direction opposite the motion. That make sense, as friction always acts to retard the motion of the object. Is friction our friend or foe? Try to walk on a perfectly frictionless surface. You simply will not get anywhere as your feet slide without propelling you forward. On the other hand, friction in a motor means that some energy might be transformed into heat or sound rather than motion. Sometimes friction is your friend and other times it is you foe. In all cases friction has to do with the relative motion of one surface upon another. We often model friction in one of two simple forms. These two forms are known as static (静摩擦 jìng mócā) and kinetic (滑动摩擦 huádòng mócā) friction. When you are pushing against an object and it does not budge, we model the force as static friction. Once the object begins to slide, the magnitude of the friction force may decrease to its kinetic value. When one surface slides over another, we usually model the motion with kinetic friction, which is also called sliding friction. Sometimes friction can depend on the speed of the relative motion. This is often the case for friction caused by motion through the air. The force of friction is related to the normal force acting on an object and can be expressed mathematically as

$$F_f = \mu N,$$

where F_f represents the magnitude of the friction force, μ is the coefficient of friction and N represents the magnitude of the normal force. In the case of static friction we can represent the coefficient of friction as μ_s and in the kinetic case we use μ_k. Suppose we consider a case similar to the one we just analyzed but now we replace the 2.00 N force with a friction force of unspecified magnitude. We also make the surfaces rough so there is some static friction between the object. Rubber on wet concrete has a coefficient of static friction of about $\mu_s = 0.300$ and a coefficient of kinetic friction $\mu_k = 0.250$. Something to keep in mind about the friction force is that it is based on the magnitude of the normal force but acts in a direction that is perpendicular to the direction of the normal force. Figure 3.3 below is similar to figure 3.2 but we have now included static friction.

The force table is now modified to include the friction force as shown below.

Will the object move with only 5.00 N applied? We can calculate the horizontal component of the applied force and compare it the friction force.

Since there is no acceleration in the y-direction, the components will sum to zero. Using the components in table 3.3 and solving for N we have,

$$N = mg - F_1 \sin\theta = 10.0\ \text{kg} \times 9.80\ \text{m s}^{-2} - 5.00\ \text{N} \sin 30.0° = 9.55 \times 10^1\ \text{N}.$$

Notice that the normal force is slightly less than the weight because of the y-component of the applied force.

The friction force (F_s) is then

$$F_s = \mu_s \times N = 0.300 \times 9.55 \times 10^1\ \text{N} = 2.87 \times 10^1\ \text{N},$$

which is much greater than the applied force of 5.00 N. The horizontal component of the 5.00 N force would be even smaller, so the block remains stationary. The friction force we calculated represents the maximum value of friction in the horizontal direction. How much force must be applied to just start the object moving? We can answer this question by finding the applied force (F_1) whose horizontal component just matches the force of static friction. For an applied force

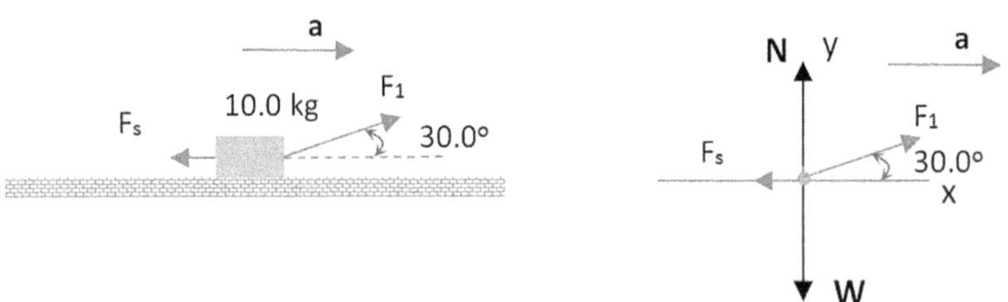

Figure 3.3. Object moving under the influence of an applied force (F_1) and kinetic friction (F_k). Physical picture (left) and free-body diagram (right). Notice that the surface has been changed to emphasize the friction force.

Table 3.3. Force components including friction.

Force	x-component	y-component
F_1	$F_1 \cos\theta$	$F_1 \sin\theta$
F_s	$-\mu_s N$	0
N	0	N
W	0	$-mg$

greater than this value we would expect the object to accelerate. Keep in mind that vertical or y-component of the applied force reduces the normal force, which decreases the friction force. Once the object does begin to move, we should change the coefficient of friction to the kinetic or sliding case. Let's calculate the magnitude of the applied force (F_1) such that its horizontal component just matches the friction force. In this case there is no acceleration. From the components in table 3.3 we have,

$$F_1 \cos\theta - \mu_s N = 0 \text{ and } N = mg - F_1 \sin\theta.$$

Notice that the normal force is a function of the applied force and we have to solve the equations simultaneously.

Solving for F_1 we obtain,

$$F_1 = \frac{\mu_s mg}{\cos\theta + \mu_s \sin\theta} = \frac{0.300 \times 10.0 \text{ kg} \times 9.80 \text{ m s}^{-2}}{\cos(30.0°) + 0.300 \sin(30.0°)} = 2.89 \times 10^1 \text{ N}.$$

If the applied force was say 35.0 N, the block would accelerate and we would have to use the coefficient of kinetic friction. In this case we cannot set the horizontal components to zero. The sum of the forces in the x-direction is the mass of the object multiplied by the acceleration. This is expressed as,

$$F_{\text{net}x} = F_1 \cos\theta - \mu_k N = ma \text{ and } N = mg - F_1 \sin\theta.$$

Solving for the acceleration and changing to $\mu_k = 0.250$ we obtain,

$$a = \frac{35.0 \text{ N} \cos(30.0°) - 0.250[10.0 \text{ kg} \times 9.80 \text{ m s}^{-2} - 35.0 \text{ N} \sin(30.0°)]}{10.0 \text{ kg}} = 1.02 \text{ m s}^{-2}.$$

In the beginning of this chapter we pointed out that many interesting devices are described in the 墨子 Mòzǐ. Canon **B26** describes the process of lifting and lowering. Translators give a number of disparate interpretations of this section. What seems clear is that it describes lifting an object by the use of a counterweight and a rope. This is illustrated in figure 3.4 shown below.

The pulley is hung from some fixed structure. A rope is attached to the load and passes over the pulley to the counterweight. This arrangement is sometimes called

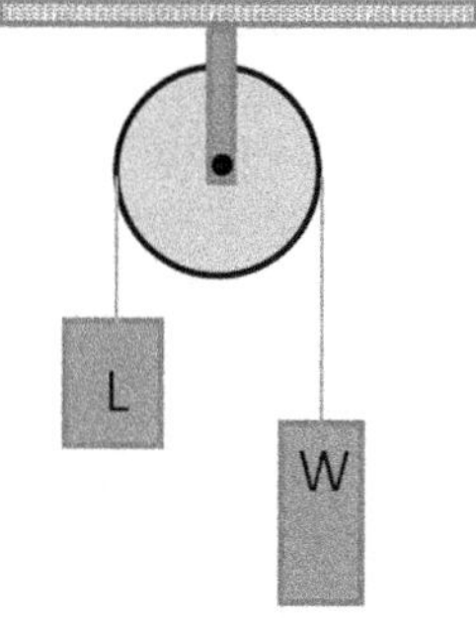

Figure 3.4. Lifting a load with a counter weight as suggested in the 墨子 Mòzǐ.

Atwood's machine in honor of George Atwood. Atwood was an English mathematician who used this machine in the late 18th century to demonstrate Newton's laws of motion.

Here are some excerpts from Johnston's translation (Johnston 2010). Canons and Explanations **B26**:

> Lifting and lowering are in opposition. The explanation lies in compelling (force).
>
> Lifting: Lifting up, there is force; drawing down, there is no force. It is not necessary that what raises it stops in action (is direct or oblique). The rope restrains raising it, like by an awl piercing it. In lifting, what is long and heavy descends, what is short and light ascends. What ascends increasingly gains, what descends increasingly loses. If the rope is straight and the counterweight and weight are alike, then it is in balance. In lowering, what ascends increasingly loses, what descends increasingly gains.

As we mentioned previously, it is very difficult to translate ancient Chinese into smooth sounding English and also capture the essential physics. Looking at the Chinese text it is clear that lifting and lowering are described using weights and that the weights are connected to a rope. An Atwood machine is not described in any detail but it does fit the general idea of what is being discussed in the text. One clue is that when the weight and counterweight are alike, the rope is straight (or vertical) and there is balance. We are also told that when the object is lifted the counterweight is heavy and descends while the lighter object ascends. While many configurations can be imagined, a simple device meeting these criteria would be Atwood's machine.

Before we undertake any mathematical analysis of a system we always try to predict the functional form of the equation we are attempting to derive and identify the important parameters. We will apply Newton's laws of motion and derive an equation for the acceleration of the weights. We have already established that when the mass of the load and the counterweight are equal, the system is in equilibrium and the acceleration should be zero. Since the objects are connected by a rope they will have the same acceleration in magnitude but in opposite directions. If the load rises at $0.10\ \mathrm{m\,s^{-2}}$, then the counterweight falls at $0.10\ \mathrm{m\,s^{-2}}$. What about the magnitude of the acceleration? Will it be just the freefall value of $= g = 9.80\ \mathrm{m\ s^{-2}}$? Neither object is in freefall as it has to drag the other object along with it. Once we have derived an equation we can test it by substituting a value of zero for one of the masses. In that case the acceleration we obtain should be the freefall value ($a = -g$). We also know that the sign of the acceleration of either weight should switch based on the magnitude of the masses. That is to say, if the mass of the counterweight (m_W) is greater than the mass of the load (m_L), the acceleration of the load should point upward (positive). However, if we make the load heavier than the counterweight, the sign should flip to negative, indicating that the load is accelerating downward. This type of physical reasoning should always be employed before starting the mathematical analysis. Mathematical errors can often be discovered with the help of units and testing the form of the equation.

The next step is to draw a free-body diagram and generate the equations of motion. Before progressing to that step, we should discuss the behavior of ropes. Whenever two objects are connected with a rope we can draw conclusions about the magnitude of their velocity or acceleration. The tension in a straight tightrope is always the same at any point along the rope. If the tension was not the same at all points along the rope, then it would loop or contort due to the unbalanced force on any segment along the rope. We also model ropes as having no mass. They serve only to provide a flexible connection between the objects. Real ropes do have mass and taking that into account leads to a more complex analysis. For our purposes we can ignore the mass of the rope as well as its strength. Beginning students often get confused about the direction of the tension along a rope. Suppose you tie a rope around an object and pull on the rope. Imagine we tie a rope around the leg of a chair and pull on the rope; with enough force, you might be able to drag the chair across the floor. Now imagine the same situation, but you push on the rope. Does pushing on the rope result in pushing the chair across the floor? No it does not. The rope just curls up or forms a loop but it does not push the chair. Hence, the age old physics aphorism 'don't push on a rope' (figure 3.5).

Now, we can make a free-body diagram of the load and the counterweight. We will assume that the load is accelerated upward, which is taken to be the positive direction. If the acceleration of the load is upward, then the acceleration of the counterweight is downward. Notice that in both cases the tension ($\boldsymbol{T}$) points upward. You might be tempted to have it point downward on the counterweight side but that would be pushing on a rope (figure 3.6).

On the load side, the tension and the acceleration both point in the same direction and must have the same mathematical sign. On the counterweight side, they should have opposite signs in the net force equation. The weight of each object is given by $\boldsymbol{F_g}$ which always points down. Since the motion is one-dimensional we will not use a force table. The net force acting on each object in the y-direction can be written as,

Figure 3.5. Pushing and pulling on a rope.

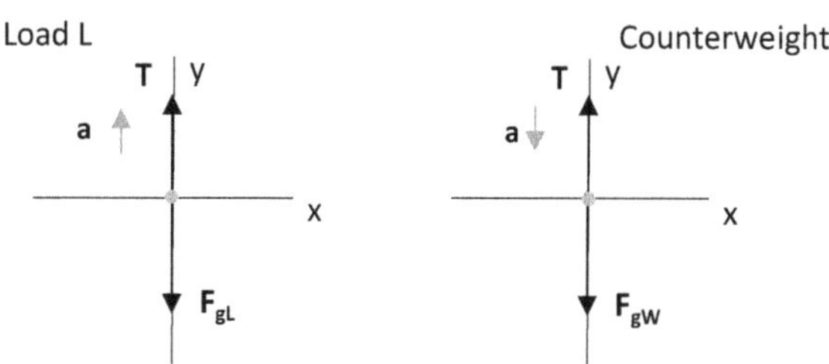

Figure 3.6. Free-body diagrams for the two weights.

$$F_{\text{net L}} = T - F_{g\text{L}} = m_{\text{L}}a \text{ and } F_{\text{net W}} = T - F_{g\text{W}} = -m_{\text{W}}a.$$

Substituting the mass and acceleration due to gravity yields,

$$F_{\text{net L}} = T - m_{\text{L}}g = m_{\text{L}}a \text{ and } F_{\text{net W}} = T - m_{\text{W}}g = -m_{\text{W}}a.$$

Notice that we now have two equations and two unknowns (T and a). Sometimes students get into the habit of thinking that the tension is equal to the weight and substitute $T = m_{\text{L}}g$. If we made such a **mistake** we would have the unphysical result that $m_{\text{L}}a = 0$ or that $a = 0$. If the object is accelerating, the tension is not simply the weight.

Solving for the tension in the equation for the load we have,

$$T = m_{\text{L}}g + m_{\text{L}}a.$$

Thus, the tension can be greater than the weight ($m_{\text{L}}g$) or less than the weight depending on the sign of the acceleration. Only when the acceleration is zero will the tension be equal to the weight. Also notice that if $a = -g$ the tension is $T = 0$ and the object is in free fall.

Finally, we substitute the result for the tension into the equation for the counterweight side and group the like terms. Solving for the acceleration we have,

$$a = \left[\frac{m_{\text{W}} - m_{\text{L}}}{m_{\text{W}} + m_{\text{L}}}\right]g.$$

Does this equation have the characteristics we predicted? Let's test it. Notice that if $m_{\text{W}} = m_{\text{L}}$, then $a = 0$ as expected. We also determined that acceleration should be less than g, the free fall acceleration. Since the term in the brackets is less than one, this prediction also holds up. We also find that if $m_{\text{L}} = 0$, then $a = g$. This makes sense since if there is no load and the counterweight just falls. Keep in mind that a points down on the counterweight side. What happens if the counterweight is missing? In that case we have $m_{\text{W}} = 0$ and $a = -g$. Remember we assumed a to be up on the load side and the negative tells us that it points down. The sign of a also depends on which mass is larger. If $m_{\text{W}} > m_{\text{L}}$, then $a > 0$ and if $m_{\text{L}} > m_{\text{W}}$, then $a < 0$.

Suppose that the load and counterweight are nearly equal,

$$m_{\text{L}} = 1.00 \text{ kg and } m_{\text{W}} = 1.10 \text{ kg}.$$

The acceleration is then,

$$a = \left[\frac{1.10 \text{ kg} - 1.00 \text{ kg}}{1.10 \text{ kg} + 1.00 \text{ kg}}\right] 9.80 \text{ m s}^{-2} = 4.67 \times 10^{-1} \text{ m s}^{-2}.$$

The equation we derived meets all of our mathematical and physical tests. This also illustrates something of the beautiful relationship between mathematics and physics. Remember that each term in an equation makes a statement about the behavior of the natural world. This also means that a mathematical error can lead to unphysical results. If you follow the methods briefly outlined in these examples you can solve many seemingly complex problems.

References

Aristotle 350 BC *Physics, Book VII* ed R Hardie and R Gaye (Transl.) Retrieved from http://classics.mit.edu/Aristotle/physics.7.vii.html

Johnston I 2010 *The Mozi, A Complete Translation* ed I Johnston (Transl.) (New York: Columbia University Press)

Newton I 1846 *Newton's Principia, The Mathematical Principles of Natural Philosophy* 1st American edn, ed A Motte (Transl.) (New York: Daniel Adee)

Chapter 4

Torque (力矩 lìjŭ)

We have briefly mentioned the topic of torque and balance when discussing the 墨子 (Mòzĭ) in chapter 2. The context was to point out that in ancient times, scientific statements were made without the use of mathematics. Some discussion of torque is also given in the experiment in which we build a 秤 chèng or steelyard balance. In this chapter we will analyze a common tool whose origins seem to point back to the ancient Chinese. The wheelbarrow is such a common device that we might just assume that it has been around as long as there have been construction projects. Needham has established that the earliest references to wheelbarrows in Europe can only be traced back to about the 13th century. Traditionally, 诸葛亮 Zhūgĕ Liàng is credited as the inventor of the wheelbarrow but there are indications it may have been invented even earlier in the first century BC (Needham and Wang 1965). 诸葛亮 Zhūgĕ Liàng is a well known figure in Chinese military history who served during the Three Kingdoms Period (三国时代 Sānguó Shídà). In 230 AD he described a device known as the Wooden Ox (木牛流马 mù niú liú mă), which could be used to carry military supplies. Regardless of the exact date of invention, the wheelbarrow, in a form we can recognize, was used by the Chinese in ancient times. Figure 4.1 shows a person using a wheelbarrow to carry a heavy load. This image dates back to a Ming Dynasty (明朝 Míngcháo) work known as 天工开物 tiān gong kāi wù, which is sometimes rendered *The Exploitation of the Works of Nature*. Compiled by 宋应星 Sòng yīng xīng, this monumental encyclopedia covered a wide range of technologies.

We can use this image to illustrate some properties of torque (力矩 lijŭ), but first we need to do some analysis. In order to analyze this picture, we need to determine if it is drawn to scale. Most people do not realize it, but the length of your foot is very nearly the distance from your wrist to elbow. This is a useful fact when analyzing images that depict a human. In figure 4.2 we have placed a red line at the foot and copied the exact line at the forearm. There seems to be reasonable agreement in the scale.

doi:10.1088/2053-2571/ab03cbch4

Figure 4.1. Early Chinese wheelbarrow. (Song, Yingxing, Tian gong kai wu, 1637, via Bibliothèque Numérique Asiatique.)

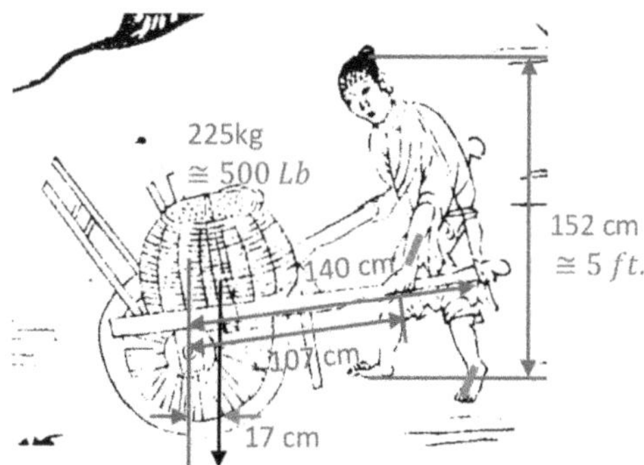

Figure 4.2. Dimensions assuming a height of 152 cm (5 ft) hunched over.

Since these distances scale correctly, we will assume that the entire drawing is to scale. Now that we have established this fact we can begin our analysis. The person is hunched over, which would take a few centimeters off of his height. We will assume, that as drawn, he stands about 152 cm (5 ft) hunched over. Now the question we want to explore is how much force must he apply in order to balance the load in the wheelbarrow? We are not concerned with forward motion or acceleration but just simply lifting the handle to a horizontal position. Notice where the load is positioned. It is not directly over the wheel nor is it in front of the wheel. If the center of mass was directly over the top of the wheel, he would only have to lift the comparatively lightweight handles. That orientation would be unstable and he could easily dump the contents over, especially on rough terrain. It makes sense to have the load located a little bit behind the axle. We also see that the wheelbarrow has struts that enable him to let go of the handles and rest. The struts are on the same side as the load, so it will tip down onto them if he lets go of the handles. His hands are not the only means of support. There is also a rope over his shoulders to help support the weight.

Torque (τ) is the product of the force and the perpendicular distance from the axis of rotation to the line of action of the force. The perpendicular distance between the axis and the line of action is called the lever arm (力臂 lìbì) or the moment arm. This is illustrated below (figure 4.3). A rod is subjected to a force F that causes it to rotate about an axis. The perpendicular distance from the axis to the line of action of the force is $r_\perp$.

This can be expressed mathematically as

$$\tau = r_\perp \times F = r \times F \sin\theta.$$

For our purposes, we will assume that the forces are applied perpendicular to the lever arm so the torque becomes

$$\tau = r \times F \sin 90^\circ = r \times F.$$

When a force causes a clockwise rotation we consider the torque produced to be negative. A positive torque corresponds to a counterclockwise rotation. We can model the wheelbarrow as a beam supported by a fulcrum (支点 zhīdiǎn) with several forces acting along the beam. When an object is in rotational equilibrium the net force and the net torque is zero. The torque model is shown below (figure 4.4). The force N is the normal force and it acts directly through the axis so its moment arm is zero.

The force supported by his hand is F_h and the force supported by his back is F_b. W_L represents the weight of the load and W_B represents the weight of the beam. Usually the weight of the beam is small compared to the load so we will neglect it.

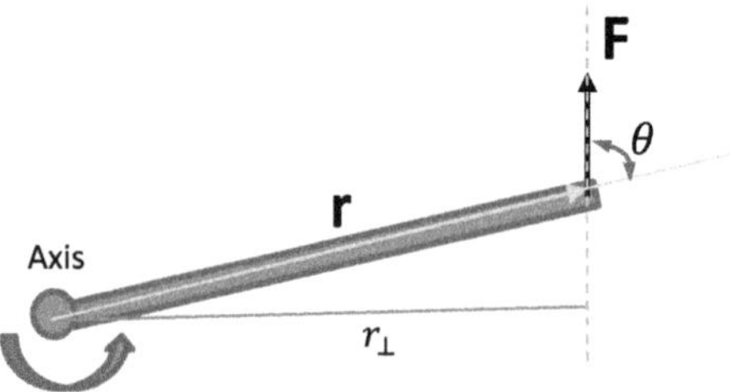

Figure 4.3. Relationship between the line of action of a force and the moment arm $r_\perp$.

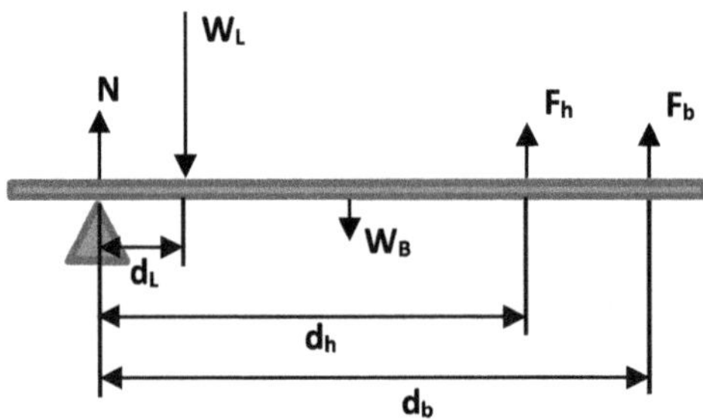

Figure 4.4. Torque model for wheelbarrow.

To calculate the net torque we must be sure to account for the sign. When thinking about the rotation direction, always imagine that the force in question acts unopposed and then consider which direction of rotation it will cause. Both F_h and F_b cause counterclockwise rotations and so contribute positive torques. W_L causes a clockwise rotation and its torque is negative. The net torque is then,

$$\Sigma\tau = 0 = -W_L \times d_L + F_h \times d_h + F_b \times d_b.$$

The weight of the load is

$$W_L = m \times g = 2.25 \times 10^2 \text{ kg} \times 9.80 \text{ m s}^{-2} = 2.21 \times 10^3 \text{ N}.$$

We will ignore the back support and solve for only hand support. In this case we can solve for F_h in the torque equation.

$$\frac{W_L \times d_L}{d_h} = F_h = \frac{2.25 \times 10^2 \text{ kg} \times 9.80 \text{ m s}^{-2} \times 0.17 \text{ m}}{1.07 \text{ m}} = 3.50 \times 10^2 \text{ N}.$$

This is equivalent to lifting only 35.7 kg. Of course, we have not taken into account that there are actually two beams or handles. If everything is symmetric, and each hand shares the load equally, it feels like lifting half of the force F_h in each hand. Now if we include the back, we have to make an assumption about F_b. You can probably lift more with your back than with your hands. Suppose you can lift twice as much with your back as you can with your hands. Then we can make the substitution

$$F_b = 2F_h,$$

which yields $W_L \times d_L = F_h \times d_h + 2F_h \times d_b$. Solving for F_h, we now have

$$\frac{W_L \times d_L}{(d_h + 2d_b)} = F_h = \frac{2.25 \times 10^2 \text{ kg} \times 9.80 \text{ m s}^{-2} \times 0.17 \text{ m}}{(1.07 \text{ m} + 2 \times 1.40 \text{ m})} = 9.69 \times 10^2 \text{ N}.$$

This is only like lifting 9.99 kg or about 20 pounds. Of course you are using your back to lift twice that force. Again, we only calculated this based on one beam but there are clearly two handles. We can also appreciate how the Wooden Ox could lead to a military advantage. Your army can carry their own provisions without the need of pack animals. This ox does not eat much!

Reference

Needham J and Wang L 1965 *Science and Civilisation in China, Physics and Physical Technology Part II, Mechanical Engineering* vol 4 (Cambridge: Cambridge University Press)

Chapter 5

Energy (能 Néng) and Momentum (动量 Dòng liàng)

We open this chapter with a drawing depicting an amazing confluence of ancient Chinese technologies (figure 5.1). In the lower left portion of the image we see what looks like some kind of gas or fire coming from a wellhead. Just above and to the left is a drilling platform. On the right-hand side we see workers in a building with many bamboo tubes. What is going on? The answer lies in simple, humble table salt. Today we have salt shakers at home and in every restaurant. Much of our fast food is excessively salty and we do not think about salt as something important, but there was a time when salt was considered to be a valuable commodity. If you are employed you should be paid a 'salary'. That word itself comes from the Latin *salarium*, which contains *sal* or salt. The reference to salt here is an allowance paid to a Roman soldier to purchase salt. Sometimes we even use the expression 'He is not worth his salt'.

Behind the workers in the picture are large vats containing salt brine. We also see some drilling derricks that remind us of oil wells. Though strange to us, this would not be an uncommon scene in 16th century 四川 Sìchuān. Around the 16th century AD, the Chinese found a way to combine salt production with natural gas. Natural gas wells, known as fire wells (火井 huǒjǐng), were drilled even earlier. There are records of natural gas wells dating back as far as 61 BC (Zhong and Huang 1997). Natural gas sometimes was found in the same area where salt brine was mined. Occasionally, even the same well would produce natural gas and salt brine. By the 16th century, during the Ming Dynasty (明朝 Míngcháo), the two products were used in a synergistic way. Natural gas from the well was burned to heat large vats of salt brine. As the water is boiled off, large cakes of salt are produced. In the center of figure 5.1 we see a tower with a worker. There is another bamboo pipe that runs into a large container that seems to feed multiple pipes over the top of the evaporating

doi:10.1088/2053-2571/ab03cbch5

Figure 5.1. Production of salt by evaporation of salt brine. (Fang, Salt evaporation in Sichuan, 1882, via Wikimedia Commons; https://commons.wikimedia.org/wiki/File:Salt_evaporation.png.)

vats. Also in the center of the image there appears to be a barrel with several pipes leading to the structure. This is most likely a device that acted as a carburetor. If you want to burn natural gas safely, you must have the correct air fuel mixture otherwise you may have an explosion. Fire wells were probably first discovered by disastrous accidents.

盐 yán is the modern character for salt. In a way, you can see something of the salt brine process reflected in the character. The top left part of the character has the radical 土 tǔ which means soil, earth. The lower radical 皿 mǐn has the meaning of a dish or shallow container. Although this might not be the official etymology of the character it does bring together the ideas. Brine and natural gas are not always found together so eventually bamboo pipelines were created as shown in this 19th-century image (figure 5.2).

In figure 5.3 we see a worker holding a percussion drill over the top of a well mouth. This appears to be a fishtail (鱼尾 yúwěi) drill bit, which had a mass of 150–250 kg and a width of 30–40 cm. Such a bit is used in the initial stages to open the well. Once the well is started, smaller drills with a mass of 100–150 kg and a width of 10–15 cm are used (Zhong and Huang 1997). In the figure you can see a small ladder. This is for the work crew who stand on the central plank. Their job is to step onto the plank, which lifts the drill bit up about 60 cm or so. The motion is something like a seesaw. There are two platforms on either side of the central plank. Workers would step off the side planks onto the lever arm. The rear of the lever arm would then strike the ground. After this the workers would step off the lever and onto the side planks, which would cause the drill to fall back down. To see a modern day recreation of the process, check the additional resource: *Machines of Ancient China (Full Documentary)* https://www.youtube.com/watch?v=XC-R4SsXRHk.

Figure 5.2. A 19th-century image of Ziliujing (自流井 Zìliújǐng), Sichuan. (The Gasser, Tzuliuching, 2013, via Wikimedia Commons; https://commons.wikimedia.org/wiki/File:Tzuliuching.png.)

Figure 5.3. Percussion drilling technique used for drilling salt wells. (Fang, Drilling a well in Sichuan, 1882, via Wikimedia Commons; https://en.wikipedia.org/wiki/Salt_in_Chinese_history#/media/File:Drilling_a_well.png.)

Over and over and over again the drill would pound the ground below. There was no shortage of labor. This process often took years but they were able to drill down hundreds of meters. When we think of drilling in the modern day we naturally think of a twist drill that rotates. The method described here is percussion drilling, which is simply a smashing and crushing process. A bamboo fiber cable is connected to the drill head and is reeled out as the drill makes its way into the ground. With this cable workers could remove the drill and lower other tools into the well. We can see this cable in the image, which is unfortunately cut off. In the original image the cable is shown wrapped around a large wheel. The animals in the picture would walk in a circle to drive the wheel and hoist the drill. After many cycles of drilling, the bore hole would fill with debris and sometimes water. The drill head was removed from the well and a bamboo tube with a valve on one end would be inserted into the well. As the tube was pushed into the well the valve opened and debris entered the tube. Pulling the tube upward allowed the weight of the contents to close the valve. This tube was then withdrawn from the well and emptied. Once the desired salt brine was reached a similar mechanism was used to remove the brine and transfer it to a storage container.

Energy Conservation (能量守恒 Néngliàngshǒuhéng)

Salt mining in China has a long history dating back over 2000 years and there are many resources detailing that history. Our objective here is to understand the drilling technique in terms of the principle of energy conservation and force.

When an object is moving it has energy by virtue of its motion. We call this kinetic energy (动能 dòngnéng), which can be expressed mathematically as

$$k = \frac{1}{2}mv^2,$$

where m is the mass of the object and v is the velocity. The units of energy are the Joule, which is expressed in units of kg m^2 s^{-2}, 1 J = 1 kg m^2 s^{-2}. We must be very careful about the units and express the mass in kg and the velocity in m s^{-1}. From the form of the equation you can see that kinetic energy is proportional to the mass but proportional to the square of the velocity. This is a very important observation. If two objects are moving at the same velocity but one object is twice the mass of the other, the larger object has twice the kinetic energy of the lighter object. However, if two objects have the same mass but one moves with twice the velocity, the faster object has four times the kinetic energy of the slower object.

Another form of energy is known as potential energy (势能 shìnéng). The name is derived from the idea that this energy is stored and has the potential to be released. There are many different forms of potential energy such as gravitational, elastic, electric, and chemical, just to name a few. We will be mainly concerned with gravitational potential energy. When you lift an object you must exert a force on the object. As you lift the object higher you must do more work against the gravitational force that wants to pull the object back down. The term 'work' is defined a little differently in physics than the way it is commonly used. In the physics definition of 'work' a force must cause a displacement or motion of the object. If there is a force but no displacement we say that no work is done. We can express work mathematically as

$$w = Fd\cos\theta,$$

where F is the force, d is the displacement and θ represents the angle between the force and the displacement as shown in figure 5.4.

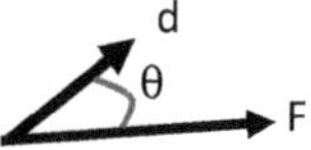

Figure 5.4. Orientation of force (F) and displacement (d) for calculating work.

Notice that the cos θ term can be positive, negative or zero. This means that depending on the relative orientation between the force and the displacement, work can be positive, negative or zero. If the force and the displacement are in exactly the same direction then the angle between the two is zero. In this case we would have

$$w = Fd\cos(0) = Fd,$$

since $\cos(0) = 1$. If the force and displacement are perpendicular then the angle is 90° and the equation becomes

$$w = Fd\cos(90°) = 0.$$

If the angle is 120°, then the work done by the force will be negative since

$$\cos(120°) < 0.$$

If an object of mass m moves upward, the gravitational force does work against the motion of the object. That means that in lifting the object we must work against gravity. In lifting the object we increase the gravitational potential energy of the object. Only changes in gravitational potential energy have physical significance. The zero point of gravitational potential energy is arbitrary and we can make it any position we wish. If we specify the zero point to be ground level and then lift the object a height h above ground level the gravitational potential energy is given by

$$U_g = mgh,$$

where g is the acceleration due to gravity and h is the height above the zero potential energy reference position. This is shown schematically in figure 5.5.

If we add the kinetic energy of an object to the potential energy, we have a quantity known as the total mechanical energy E_{mech}.

$$E_{\text{mech}} = U + k.$$

Here the U can represent any potential energy such as elastic potential or gravitational potential energy.

There is a very important principle in physics that says in an isolated system where only conservative forces act, the total mechanical energy of the system cannot change. We call this principle the conservation of mechanical energy.

Energy can change form but the total must remain the same. We can now apply this concept to the salt well drill and perform a few calculations.

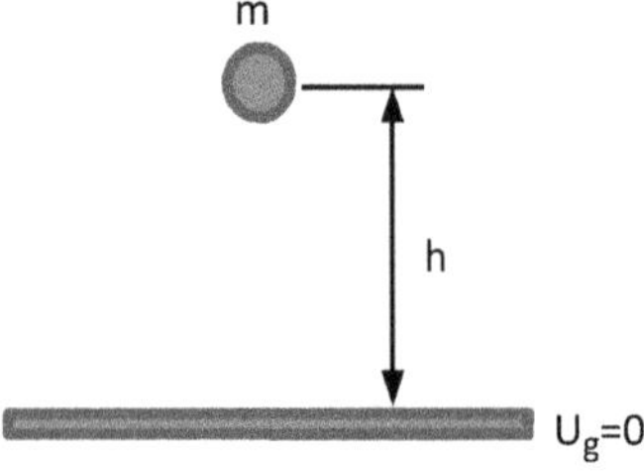

Figure 5.5. Gravitational potential energy of an object mass m at a height h above the reference position.

As mentioned previously, the mass of the fishtail bit was as much as 2.5×10^2 kg so we will use this in the calculation. In each cycle the bit is lifted by only about 6.0×10^1 cm. From modern recreations it takes about 5.0 s for each cycle. These are only approximate numbers so we will use only two significant figures for the calculation.

We can calculate the gravitational potential energy of the drill bit, assuming the start of its journey is the zero of potential energy.

$$U_g = mgh = 2.5 \times 10^2 \text{ kg} \times 9.80 \text{ m s}^{-2} \times 0.60 \text{ m} = 1.47 \times 10^3 \text{ J} = 1.5 \times 10^3 \text{ J}.$$

We will use these numbers for multiple steps and taking 1 more significant figure is permitted for the intermediate steps but not the final results. If we were to round at each step, the rounding errors would compound.

If this drill bit falls without friction from the mechanism, then all of that potential energy would be converted into kinetic energy. In that case we can calculate the velocity of the drill bit when it strikes the ground. In terms of the total mechanical energy we can write

$$E_{\text{mech}i} = E_{\text{mech}f},$$

where 'i' means initial when the drill is raised above the ground, and 'f' means final when it just strikes the ground. In terms of the kinetic and potential energies we have

$$(K + U_g)_i = (K + U_g)_f.$$

We let U_{gf} be our reference position so $U_{gf} = 0$. We can then substitute our equations for the kinetic and potential energies and obtain

$$\frac{1}{2}mv_i^2 + mgh_i = \frac{1}{2}mv_f^2.$$

Now, if the drill starts from rest the initial velocity is zero so $v_i = 0$ and we have

$$mgh_i = \frac{1}{2}mv_f^2.$$

We can solve this equation for v_f, which is the velocity the bit has when it just contacts the ground. Notice that the mass is the same on both sides of the equation so that divides out and the velocity only depends on the height. But we know that being struck by a heavy object falling from 1 m is very different from being struck by a lightweight object falling from the same height. This is true and we shall discuss this in the section on momentum. Solving for the velocity we have

$$v_f = \sqrt{2gh_i} = \sqrt{2 \times 9.80 \text{ m s}^{-2} \times 0.60 \text{ m}} = 3.4 \text{ m s}^{-1}.$$

Each time the drill bit is raised and lowered, energy is transferred from gravitational potential energy into kinetic energy. This is repeated over and over again for years. The rate at which energy is transferred or the work is done is called power (功率 gōnglǜ). Power has units of Joules per second, which are also called Watts ($1 \text{ J s}^{-1} = 1 \text{ W}$). The name comes from James Watt who did a great deal of research on steam engines. We can express the power mathematically as

$$P = \frac{\Delta E}{\Delta t},$$

where ΔE is the amount of energy transferred and Δt is the time interval. If we take our 1.47×10^3 J of energy and transfer that in an interval of 5.0 s, this corresponds to a power output of

$$P = \frac{\Delta E}{\Delta t} = \frac{1.47 \times 10^3 \text{ J}}{5.0 \text{ s}} = 2.94 \times 10^2 \text{ W} = 2.9 \times 10^2 \text{ W}.$$

This might not seem like much but if the drill runs continuously for a year, then a great deal of energy is transferred. Just for the sake of calculation let us assume that it does transfer energy at this rate for one year continuously. This is somewhat unrealistic since the well had to be cleaned and there were mechanical failures but we will do the calculation for practice.

If we multiply power (P) by time (Δt) we get back to energy

$$P \times \Delta t = \frac{\Delta E}{\Delta t} \times \Delta t = \Delta E.$$

The time interval we are interested in is one year so we can calculate the number of seconds in a year as

$$1 \text{ yr} = 365.25 \text{ day} \times 24 \text{ h day}^{-1} \times 3600 \text{ s h}^{-1} = 3.16 \times 10^7 \text{ s}.$$

So with these assumptions the total energy transferred is

$$2.94 \times 10^2 \text{ J s}^{-1} \times 3.16 \times 10^7 \text{ s} = 9.29 \times 10^9 \text{ J} = 9.3 \times 10^9 \text{ J}.$$

Energies in the Giga Joule range ($1 \text{ GJ} = 1 \times 10^9$ J) are comparable to explosions of TNT. 1000 kg or 1 tTNT has an explosive yield of 4.184×10^9 J. So in terms of tons of TNT we have

$$9.27 \times 10^9 \text{ J} \times \frac{1 \text{ tTNT}}{4.184 \times 10^9 \text{ J}} = 2.2 \text{ tTNT}.$$

One would not actually drill a well this way. The explosion would blast in all directions and not actually form a bore hole. Considering the explosive nature of a fire well this would be even more dangerous. We are just comparing energy output.

Momentum and Force

We previously noted that a heavy object moving at a few meters per second seems to produce a greater impact force than a lightweight object moving at the same velocity. This is related to the concept of momentum. Mathematically, momentum (P) is the product of the mass of the object with the velocity, which can be written as

$$P = mV.$$

To change the momentum one must apply a force. A collision is a short time event that causes a change in momentum. The force required to change the momentum of an object is related to the rate at which the change occurs. The more abrupt the change the greater the force that is required. This relationship is expressed mathematically as

$$F_{\text{avg}} = \frac{\Delta P}{\Delta t},$$

where ΔP is the change in momentum and Δt is the time interval during which it changes. Here, we expressed the force as the average force during the collision or impact. The quantity ΔP is also known as the impulse. When the percussion drill bit strikes the rock it is brought to rest very sharply. The force is exerted by the rock to bring the drill bit to rest. Newton's law tells us that this is the same as the force exerted on the rock by the bit. To calculate the change in momentum we use

$$\Delta P = P_f - P_i,$$

where $P_f = 0$ when the object stops moving. For our 2.5×10^2 kg drill bit moving at 3.43 m s^{-1} we have an initial momentum of

$$P_i = -2.5 \times 10^2 \text{ kg} \times 3.4 \text{ m s}^{-1} = -8.6 \times 10^2 \text{ kg m s}^{-1}.$$

Momentum is a vector and has a direction associated with it. The initial momentum is downward, which we take to be in the negative direction.

The magnitude of the average force exerted on the rock now depends on the timescale of the collision. We do not know what that contact time is. A simple experiment with a masonry chisel is shown in figure 5.6. The sensor is a piezoelectric transducer that produces a voltage when a force is applied. The vibration in the crystal takes tens of milliseconds to die out. From the response, we estimate the duration of the initial impact to be 2.08 ms. How well this matches the actual drill used in ancient times is not known. A result on the order of a few milliseconds seems

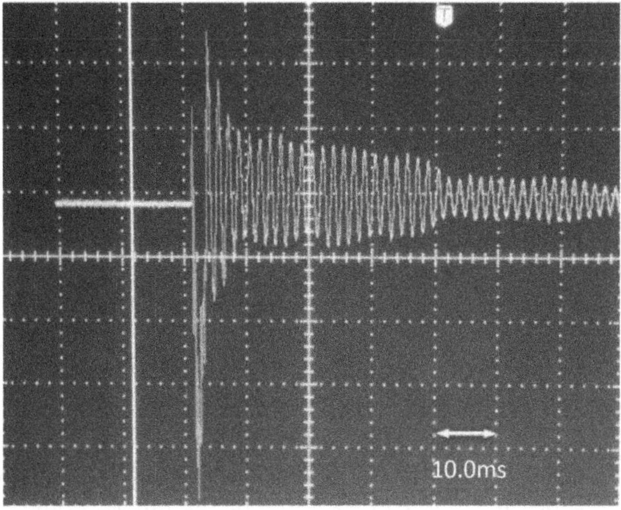

Figure 5.6. Impact of a masonry chisel measured with a piezoelectric sensor.

reasonable for contact between two hard materials and we will use this in our calculation.

$$F_{avg} = \frac{\Delta P}{\Delta t} = \frac{0 - (-8.6 \times 10^2 \text{ kg m s}^{-1})}{2.08 \times 10^{-3} \text{ s}} = 4.1 \times 10^5 \text{ kg m s}^{-2} = 4.1 \times 10^5 \text{ N}.$$

This is only the average force and the peak force may even be greater, it all depends on the shape of the impulse curve, which we do not actually know.

While the force we estimated is very impressive, we have to consider how that force is imparted to the rock. In the introduction to this chapter we said that the fishtail drill was 30–40 cm wide. The masonry chisel used in this experiment has a thickness of 1.42 mm at the sharp tip. Pressing against a needle with 1 N of force feels very different than pressing on a flat surface with the same force. The important factor here is the contact area. We will make one more assumption, this time about the contact area of the drill bit and calculate the pressure exerted on the rock.

Suppose the thickness of the edge of the drill bit is about the same as the masonry chisel and the width is 40 cm.

The contact area would be

$$a = 4.0 \times 10^{-1} \text{ m} \times 1.4 \times 10^{-3} \text{ m} = 5.6 \times 10^{-4} \text{ m}^2.$$

Pressure is defined as the force per unit area and has units of N m^{-2}. The pressure unit is also known as the Pascal (Pa) with 1 Pa = 1 N m^{-2}. The pressure exerted on the rock would be approximately

$$P = \frac{4.1 \times 10^5 \text{ N}}{5.6 \times 10^{-4} \text{ m}^2} = 7.3 \times 10^8 \text{ Pa} = 7.3 \times 10^2 \text{ MPa}.$$

The unconfined compressive strength of a rock has a wide range. Hard rocks, like granite, can be as high as 3.45×10^2 MPa, while softer sandstones can be as low as 4.0×10^1 MPa (Hoek and Brown 1980). From all the assumptions we made, we have calculated that the drill head could apply about twice the pressure required to break granite. In reality, the pressure produced was likely to be much lower. As you can see, we are only making educated guesses about the contact area and the impact time. We did not make any assumption about the material properties of the drill bit. Repeated impacts would flatten the sharp edge of the bit. This, in turn, increases the contact area and decreases the pressure. However, considering all the assumptions we made in these calculations our numbers look reasonable to within the correct order of magnitude, which is about the best we can hope for.

The equation for average force might save your life someday.

$$F_{avg} = \frac{\Delta P}{\Delta t}$$

gives the basic physical principle behind air bags. In an automobile accident your body is moving at the same speed as the car. In a crash you fly forward and strike something fixed such as the steering wheel. If you struck the steering wheel abruptly, the Δt would be short and the average force of the impact could be dangerously high. The soft cushion of the airbag extends the time it takes you to come to rest and thus, decreases the force of the impact. Physics is all around us hiding in plain sight.

References

Hoek E and Brown E T 1980 Empirical strength criterion for rock masses *ASCE J. Geotech. Eng. Div.* **106** 1017

Zhong C and Huang J 1997 *Drilling and Gas Recovery in Ancient China*

Chapter 6

Experiments

In this section we provide several laboratory experiments to complement the material in the preceding chapters. All of these experiments have been designed with the twofold objective that permeates this entire book. First we look at an invention or technology from ancient China. Each experiment introduces related Chinese vocabulary and a short historical narrative. Then there is a detailed procedure describing how students can recreate the invention or technology using modern materials. The materials and equipment were chosen with consideration to cost and availability. In some cases, simple hand tools such as a hacksaw or drill are required. In the case of paper making, entire paper-making kits are available and recommended sources are provided. Such sources may change in time so it is recommended that the laboratory instructor does a little research on their own. The second layer of the twofold approach is to analyze what has been constructed. For the purposes of a laboratory science class, it is this analysis that brings in applications of the scientific method and mathematics. Of course, the laboratory instructors are encouraged to innovate and adapt the experiments to suit the needs of their particular students. Safety has also been an important consideration in the design of these experiments. When working with tools there is always the chance of a cut or some other injury. Depending on the age group, it may be prudent for the laboratory instructor to pre-cut some components.

Most of the experiments described require more than one laboratory period, based on a three-hour lab period commonly found at universities. Graphs are a powerful tool and we try to incorporate graphical analysis in each laboratory experiment. Over the years of performing these experiments, students often comment on the new hands-on skills that they acquire. Building something from 'scratch' is a lost art in a world of computer simulations and simulated laboratory experiments. Making something with their own hands helps to connect students with the ancient craftsman and artisan whose world we seek to understand. Students also

doi:10.1088/2053-2571/ab03cbch6

take great pride in having an artifact to remind them of their laboratory experience and to show others what they have accomplished.

List of laboratory experiments

- Paper making: measurement techniques and graphical analysis.
- Silk: stress–strain and elasticity.
- Cheng balance: torque and rotational equilibrium.

6.1 Paper making (造纸 Zàozhǐ)

Objective

This experiment consists of two parts and will take several laboratory sessions to perform. In the first part we will learn about paper making and practice making a few sheets. In the second part of the experiment we will analyze our paper and the paper-making process. Measurement is a key idea in science. Without measurements and a method to analyze those measurements, how can one do science? Graphical analysis is an important scientific tool that helps us make sense and extract meaning from our measurements. Making a sheet of paper is rather simple to do but explaining it to others requires attention to detail. In reporting scientific results it is important to describe the method used so that it can be reproduced by others. This experiment then, is not just the simple act of making paper, but rather an introduction to scientific methods and laboratory techniques.

Chinese vocabulary

纸 zhǐ paper	造 Zào to make, build, invent, fabricate
纸浆 zhǐjiāng paper pulp	浆 jiāng any thick fluid
水 shuǐ water, liquid	蔡伦 Cài Lún the inventor of paper

Materials

Deckle (wood frame)	Sponge	Plastic measuring cup (1 L)
Paper-making screen	Blow dryer	Pressing block
Cover screen	Mass scale	Press weight (several kg)
Screen support grid	Iron and ironing board	Dial indicator and base or caliper
Vat for pulp (5 L or more)	Blender	Gloves
Cookie sheet	Fiber source or scrap paper	Drying sheet

Paper-making kits can be found at Arnold Grummer's web site, https://arnoldgrummer.com.

Introduction

What is paper? You might think that the answer to this question is obvious but people do find it difficult to explain what paper it. Many people are under the impression that paper was invented by ancient Egyptians. That is because they think papyrus is an early form of paper. Papyrus and paper are actually rather different materials. Papyrus is made of a soft material found inside of reeds. This soft spongy material is often called pith. To make papyrus one splits the reed and removes the pith. The pith can then be pressed and smoothed to make a writing material. This material was used extensively in the ancient world and much of what we know as a civilization was recorded on papyrus strips. The Great Library of Alexandria held thousands of documents written on this material. As important as papyrus was, it is not paper. Another writing material that is often thought of as paper is parchment.

Parchment is an animal hide often made from calfskin, goatskin or sheepskin. Parchment was probably developed around 200 BC as a result of an embargo on the shipment of papyrus. People wrote on animal skins prior to the invention of parchment, but parchment had some desirable qualities that made it different than earlier skin-based materials.

So what is paper? Paper is best described as the sediment of disintegrated fibers. These fibers are obtained from various types of plant materials. Fibers may come from a wide range of starting materials including: wood, tree bark, plant stems, leaves and cotton. The defining difference is that paper is fibrous sediment. Paper usually consists of a mixture of fibers that provides particular properties. Hardwoods and softwoods can be mixed together to produce materials with varying rigidity. Sometimes linen is added to wood pulp to produce a softer paper. Blending and processing techniques are an entire scientific field of study. In the second part of this laboratory activity, you will be free to explore this area by making your own 'concoctions'.

The oldest surviving piece of paper can be dated between the years 140 to 87 BC. This paper was found near the ancient city of 西安 Xī'ān. 西安 Xī'ān is of course well known as the location of the famous Terracotta Warriors. These dates place paper production in the Han Dynasty (汉朝 Hàn Cháo). You might think that paper was immediately used for writing but it is more likely that paper was first used for wrapping or protecting precious material just like our modern packing paper. Early paper was rather rough, not like the fine milled paper we have today. Paper was also used in China for a wide range of uses that had nothing to do with writing. Some of the earliest known uses of paper were for insulation, clothing, armor, padding, artwork and personal hygiene. The invention of paper in its sediment form is credited to 蔡伦 Cài Lún, who was a Han dynasty official. As with most ancient inventions, the exact inventor might be questionable and there was no patent office. Still, 蔡伦 Cài Lún is traditionally regarded as the inventor of paper, but even he admits that he was inspired by bees and wasps. This, by the way, is an example of what modern scientists and engineers call biomimicry. That is, looking to the natural world for the solution of engineering problems. In China pre-paper writing materials consisted of bamboo strips and silk. These were not particularly convenient materials. Bamboo was heavy and cumbersome to store. Silk was easy to store but very expensive. The Chinese character for paper is 纸 zhǐ. It is interesting to note that the radical on the left side is silk. Paper was inexpensive to produce and could be stacked, folded or bound in a manner that made it easy to store documents.

To make paper we need a source of fibers suspended in water, which is called pulp. We can make our paper pulp by either recycling paper or starting with raw biomaterials such as wood chips or cotton linter. There are two important components in the pulp. These are lignin and cellulose. Lignin is a biopolymer that is found in plant material. It often gives strength and rigidity to plants and helps to transport water in plant stems. It is not, however, a desirable component in paper. Lignin degrades paper and causes it to become yellow and brittle over time. The other important material in paper pulp is cellulose. Cellulose is a main component in the cell wall of green plants. Cotton fibers are the purest naturally occurring form of cellulose. One way to think about the lignin–cellulose relationship is to think of the lignin as glue that holds the cellulose fibers together. In paper making, one tries to separate the lignin from the cellulose and use mostly cellulose. This can often be accomplished by chemical means, by attacking the lignin but leaving the cellulose undamaged. When we work with recycled paper, much of that work has been done for us. If we work with raw wood products we either have to remove the lignin or settle for making paper with lignin in it. Softwood and hardwood trees have differing amounts of lignin and cellulose. A softwood tree such as a pine fir or spruce has about 40% cellulose and about 28% lignin. Hardwoods have about 45% cellulose and 20% lignin content. Another difference is in the fiber lengths. Many softwoods have fibers that are several millimeters long while hardwood fibers range from a little under one millimeter to slightly more than a millimeter. A smooth surface for writing can be obtained from the short fibers characteristic of hardwood trees. Such fibers are also easier to bleach if you wish to make white paper.

You can make an almost cloth-like paper from cotton linter. Cotton linters are fibrous hairs found on cotton seeds. Many people regard paper made from cotton linter as superior in strength, texture and snowy white in color. You can add about 10% cotton linter to recycled paper or make paper only from cotton linter. The cloth-like feel of currency comes from cotton linter.

Experimental

In our experiments we will first start with recycled paper. Since we are starting with a paper source, much of the chemical processing has been done for us. We only need to disentangle the fibers by converting the paper back into pulp. Since there are a few techniques involved in paper making it is much easier to start with recycling. This will give you the opportunity to learn how to use the equipment and practice all the required steps. Once you have mastered all the basic steps, you can become more analytical.

The basic procedure for making a sheet of paper is as follows;

(1) For each 8 1/2 × 11 in. sheet of paper you will need approximately 4 cups (950 ml) of water. You will need to fill the blender with pulp three or four times to have enough pulp for the experiment. Record the blender setting and speed.

(2) Assemble the mold as shown below in figure 6.1. The parts are offset to show each section, but yours should be aligned with the deckle (wooden frame) on the top and the plastic support grid on the bottom.

(3) Use your hand to agitate the pulp so that it is more uniformly distributed in the vat. It is very important to have a uniform distribution of pulp. Gloves should be worn. Old bits of recycled paper may very well have bacteria on them and it is a wise precaution to use gloves.

(4) Place the mold assembly into the pulp vat with the mold initially vertical. Rotate the mold into the pulp so the top goes under the pulp last, as shown below in figure 6.2.

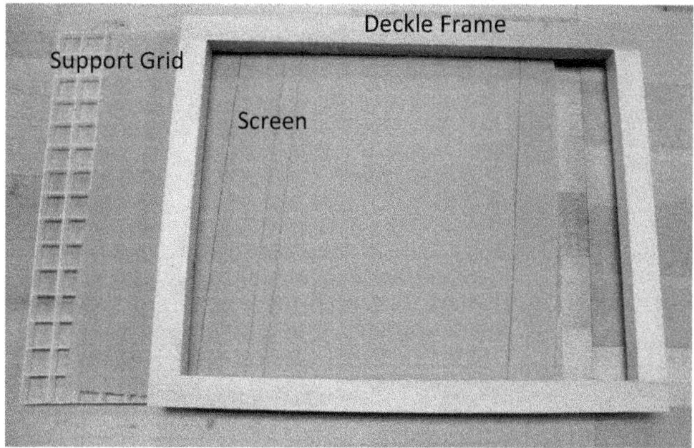

Figure 6.1. Assembling the mold.

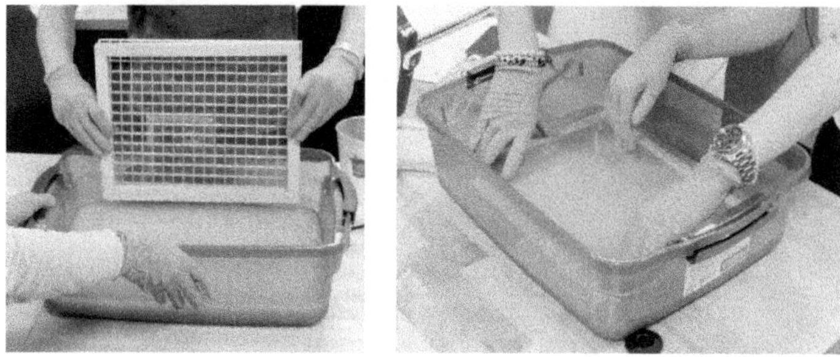

Figure 6.2. Putting the mold into the pulp.

(5) Allow the pulp to settle on the screen and then lift the mold straight up out of the pulp vat (figure 6.3).
(6) Place the mold in the vat lid or cookie sheet and allow the water to drain. Remove the deckle and you should have a wet sheet of paper on the screen as shown below (figure 6.4).
(7) Place the gray cover screen over the pulp and press dry with a sponge (figure 6.5). Wring out the sponge and return the water to the pulp vat.
(8) Peel off the cover screen and replace it with a drying sheet (figure 6.6). Sponge dry.
(9) Now flip the sheet over and slowly peel back the screen exposing the pulp (figure 6.7). Be very careful not to damage the moist pulp.
(10) Replace the screen with another drying sheet and sponge dry using some firm pressure. A press bar can be used to exert pressure on the drying sheet to squeeze out water.
(11) Now it is time to dry the paper between the two drying sheets. Use an iron and ironing board to dry the paper (figure 6.8). Ironing also keeps the paper flat as it dries out.

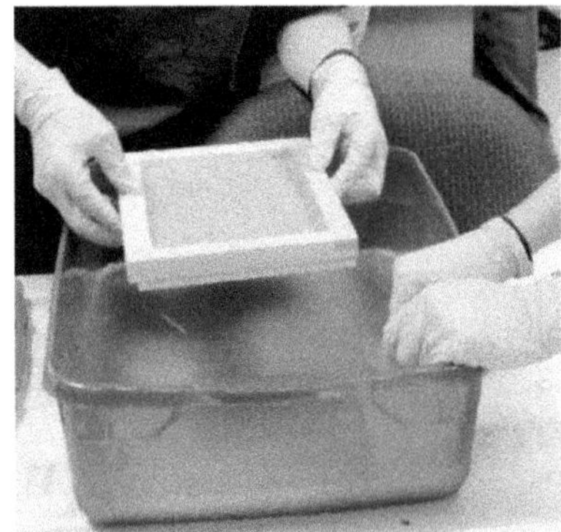

Figure 6.3. Pulp on the screen.

Figure 6.4. Paper on the screen.

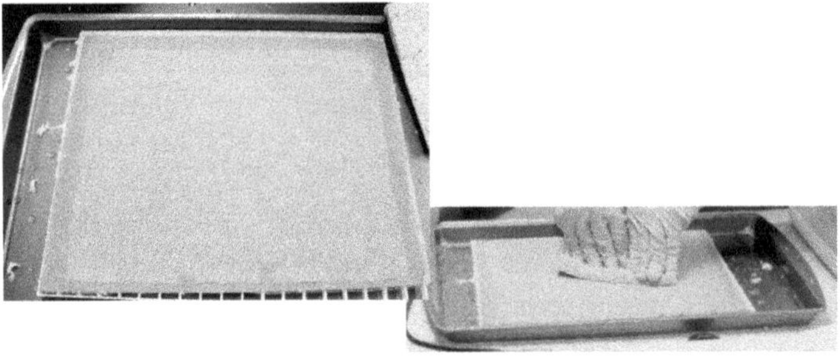

Figure 6.5. Drying the paper.

As the paper dries, you can remove the drying sheet and iron directly onto the paper (figure 6.9).

For the first paper-making experiment you need to master the technique. Once you feel you have mastered the basic technique, experiment with your own pulp concoctions. You can also explore different types of pulp, coloring agents, or

Figure 6.6. Using the drying sheet.

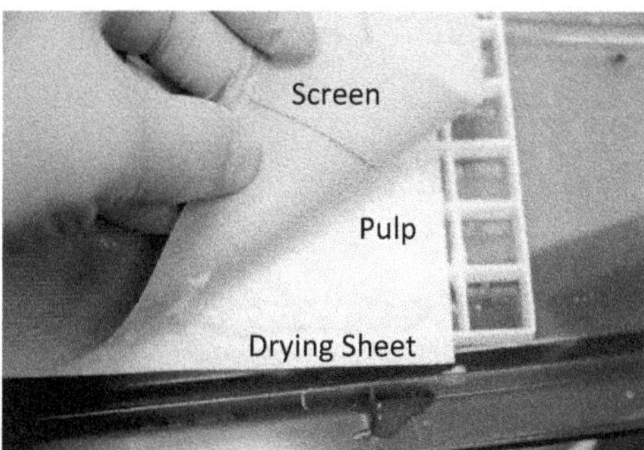

Figure 6.7. Remove the screen.

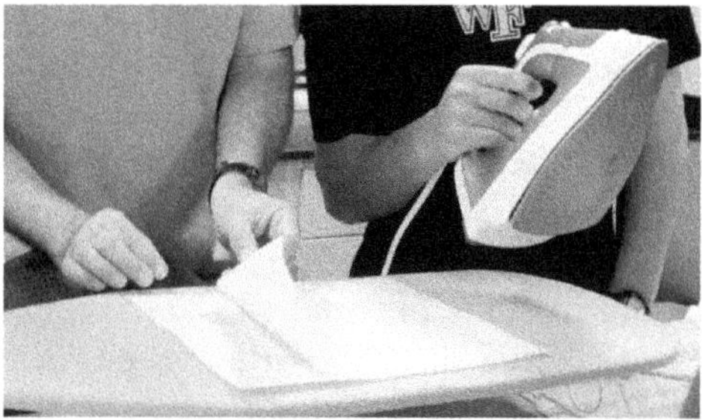

Figure 6.8. Iron the paper.

embossing. Record what you did in your laboratory notes. Include data such as the volume of water used, amount of recycled paper or other fiber sources, blender speed and blending time. It is important that you learn to document your work. Learn to write down every detail, even what does not work.

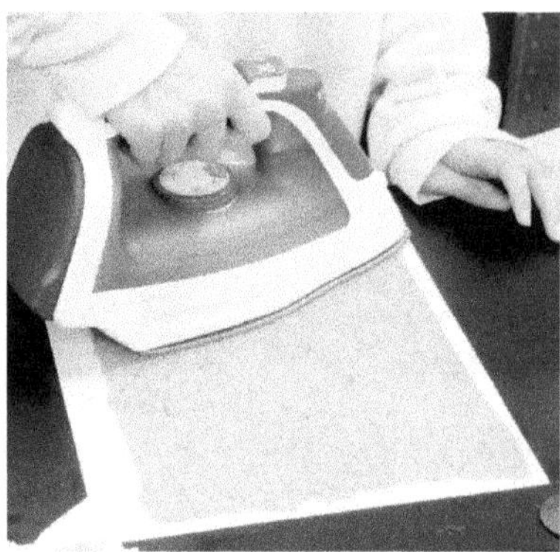

Figure 6.9. Iron directly onto the paper.

Figure 6.10. Pulp on the mold.

Paper making part 2

Analytical method

Now it is time to get a little more scientific. We want to explore some of the properties of our handmade paper. These experiments will require recording data, data analysis, and graphing. As you have seen in the lab, paper pulp does not settle uniformly (figure 6.10).

This means that a sheet of paper will have variations in thickness. We will explore some of the factors that influence the thickness. In our first experiment we applied pressure to the sponge when drying our paper. When the pulp is still wet, its thickness can easily be altered by applying pressure. When making paper by hand everyone will use different amounts of force on their sheet or even vary the force applied to different regions of the sheet. We are going to explore how the applied force alters the thickness. Not only does the thickness change, but we would expect the variations in thickness to be altered by the applied force. Pressing down hard can make the paper smoother.

Varying the drying force

Normally, we press rather hard to remove the water. In this experiment we will press with different amounts of force in three different areas of the paper. We can vary the force by simply placing a weight on top of the sponge. To apply the least amount of force we will simply lay the press block on top of the drying sponge. We can increase the drying force by adding weights to the press bar as shown below. We can divide the sheet into three sections and do the initial drying by adding weight to the press bar as we move across the paper. When the paper is wet, we cannot write on it so we can mark the sections by simply tearing the edge of the paper as shown below, see figure 6.11. When the paper is dry we will be able to identify each region.

In the section marked 'minimal force' just place the press bar on top of the sponge to give it a little force a shown below, figure 6.12.

Let the sponge remain on the drying screen for a few tens of seconds and then wring it out. Be sure to wring out the sponge over the pulp vat. We want to return the water to the pulp supply.

Once you have removed as much water as possible from the 'minimal force' region, dry the middle section. To dry this section add some weight to the press bar and work your way across the next region of the sheet. A 1.0 kg mass has been placed on top of the press bar to increase the drying force (figure 6.13).

For the final section of the paper you need to apply a large force. Use your hand to press down on the bar. Lean hard against the bar and apply as much force as possible.

In our previous procedure we used an iron to dry the sheet of paper. If we did that with this paper we would alter the results of varying the drying force. This time we will use a blow dryer to dry out the remaining water. Gently dry the paper between

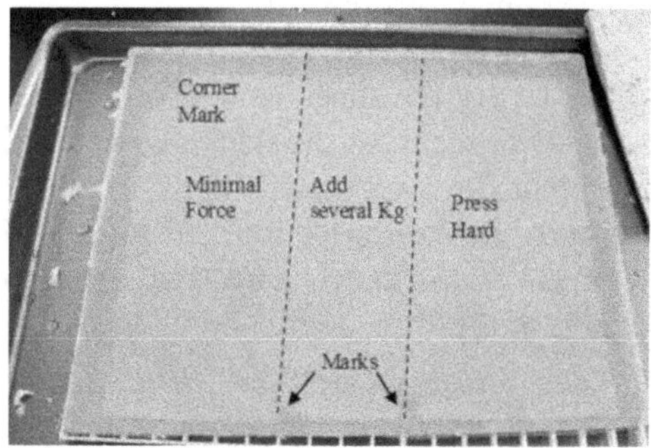

Figure 6.11. Tears on the paper.

Figure 6.12. Press bar on the sponge.

Figure 6.13. Add 1.0 kg mass to the press bar.

two drying sheets so you do not wrinkle or tear the wet sheet. Be sure to keep the blow dryer moving and try to dry the paper uniformly. Drying with a blow dryer takes longer, so be patient.

Procedure for paper making, data sheet—force variation

(1) Prepare about 6 L of paper pulp; be sure to record the paper/fiber mass, volume of water and blender speed.
(2) Measure the density of the pulp before making a sheet.
(3) Make a sheet of paper using our standard method.
(4) Tear off a corner to mark the minimal force region.
(5) Tear off the edges to indicate the areas subjected to different forces.
(6) Dry the paper with a sponge and weights as described in the text. Be sure to wring out the sponge into the vat.
(7) Transfer the paper to the drying sheets.
(8) Use a blow dryer to dry the sheet.
(9) Use a dial indicator to measure the thickness at 30 different points in each of the three different force regions.
(10) Enter this data into the data table marked **force variation**.
(11) Calculate the average thickness, standard deviation and coefficient of variation for each of the three regions.

Measuring the thickness

We can measure the thickness of our paper by using a dial indicator as shown below (figure 6.14). A Vernier caliper can be used for these measurements if a dial indicator is not available. Use a caliper or dial indicator that can resolve 0.01 mm. These instructions assume that a dial indicator is used. The indicator has a small plunger that is placed in contact with the paper. As this plunger moves up and down, its displacement is shown on the dial. The dial indicator should be mounted to a stable base or clamped to a ring stand.

First, we must set the dial to zero by allowing the plunger to touch the surface of the table. The outer ring of the indicator can be rotated so that the dial reads zero with no paper under the plunger. Once the indicator is set to zero you can slide the paper under the plunger and measure the thickness. The plunger is spring-loaded and will press the paper against the table top. When you measure the paper, tap the table a few times to help the plunger settle. As you tap the table you may notice the

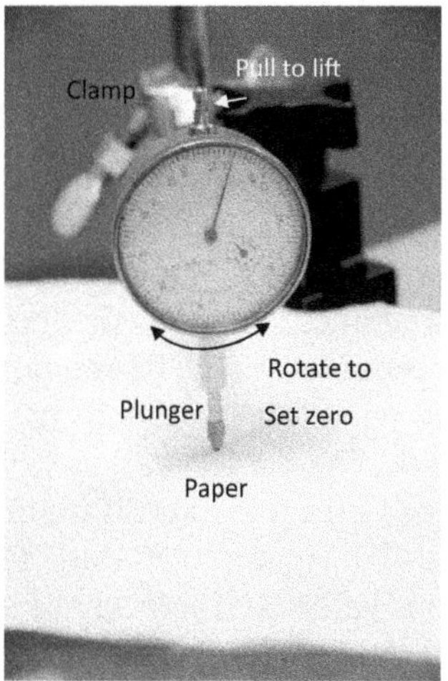

Figure 6.14. Dial indicator.

needle move. You will need to make at least **30 thickness measurements for each of the three sections** of the sheet. At the position of each thickness measurement write the data point number directly on the paper. The thickness measurements should not be made in a random order. Make a series of measurements in each of the three different sections. Placing your measurements in a predictable order will make analysis easier and help you to spot trends. For example, you organize your measurements in stripes from the top of the page to the bottom of the page. See the analysis section for alternative ways to analyze the data. Record your measurements in the data table marked Force Variation. Once you have marked all your data points hold the paper up to a light source and take a photograph. You should be able to identify thin and thick spots by how much light is transmitted through the paper. Notice that the dial is calibrated in 0.01 mm increments. This means that a reading of 30 corresponds to 30×0.01 mm = 0.30 mm.

Analysis

This experiment generates quite a bit of data. Now we can do some statistics on that data. We are interested in finding the mean thickness for each section. Remember that our objective is to determine if varying the drying force made any difference in the thickness or smoothness of the paper. This is the idea that you should discuss in your lab report. The drying force might influence the thickness in two different ways. Different regions of the paper might have a different average thickness and they might also have a smoother surface. A smoother surface means that there will be less variation in the thickness readings compared to a rough finish. We need a way to quantify just how smooth the surface is. Quantifying the variation in the thickness requires using a mathematical function known as the standard deviation (σ). The standard deviation tells us how wide the variation is about the mean value. This is written mathematically as

$$\sigma = \sqrt{\frac{\sum(x - \bar{x})^2}{(n - 1)}},$$

where $\bar{x}$ is the mean value. There are several ways to understand what the standard deviation tells us. We might just simply have random variations in the measured thickness. If that were the case those variations might follow a normal distribution. The normal distribution is sometimes known as the 'bell curve' because the shape of the probability distribution resembles a bell, as shown in figure 6.15.

Now let's try to understand what this graph tells us. Suppose we made a series of measurements of the thickness. The values we get will spread over a range where some spots may be very thin and others thicker. If we find the average or mean value of our thickness samples it will fall right at 0.50 mm. There may not actually be any measurement that is exactly this value but there will be a spread around that value. If distribution of values follows the normal distribution function then approximately 68% of our thickness measurements will fall within about 1 standard deviation of the mean value. The 68% rule works best for a very large number of samples and may not accurately represent the data with only a few tens of samples. Still, it is a useful measure for the purpose of this experiment. Since the mean is 0.50 mm and the standard deviation is 0.10 mm, about 68% of our values (a little more than half of them) will fall between a thickness of 0.50 mm − 0.10 mm = 0.40 mm and 0.50 mm + 0.10 mm = 0.60 mm. These bounds are indicated by the arrows on the graph. If a second sheet of paper still has the same mean thickness but is slightly more uniform, then the standard deviation will be smaller, as shown in figure 6.16.

For the second sheet of paper the standard deviation is smaller ($\sigma_2 = 0.06$ mm) but the mean thickness is the same. Now, approximately 68% of the measured values fall

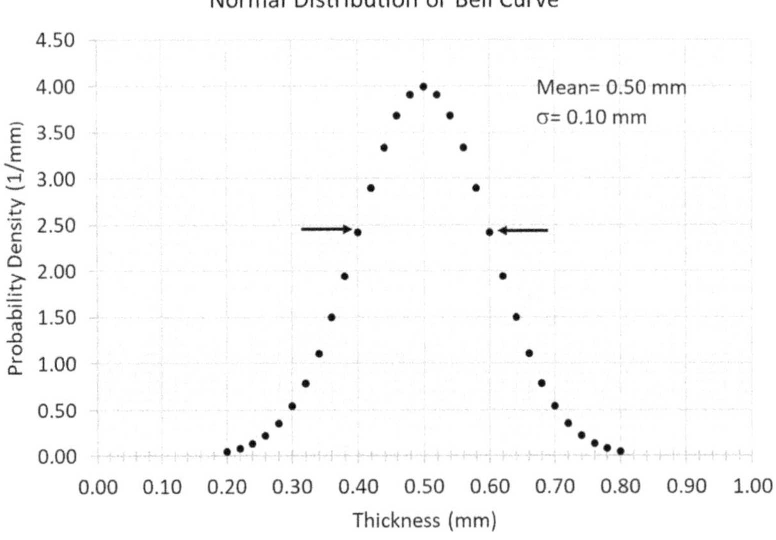

Figure 6.15. Normal distribution function also known as a bell curve. Mean thickness 0.50 mm standard deviation 0.10 mm.

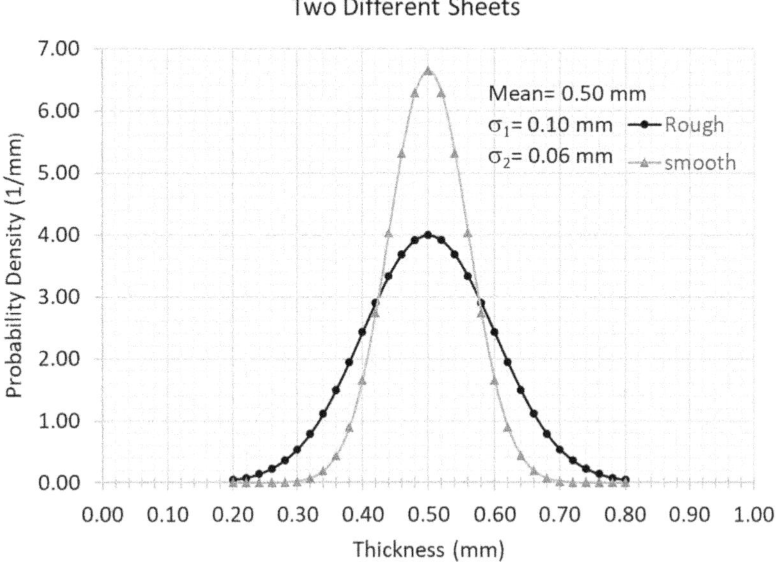

Figure 6.16. Two sheets of paper with the same mean thickness but one has less variation as shown by the smaller standard deviation.

between 0.5 mm – 0.06 mm = 0.44 mm and 0.50 mm + 0.06 mm = 0.56 mm. You can calculate the standard deviation of a data set using spreadsheet software such as Excel. In Excel the syntax for this function is

=STDEV(start cell: end cell).

For example, =STDEV(S7:S16) means find the standard deviation for data in a list starting at cell S7 and ending at cell S16.

In reality, there is no reason to expect the thickness values to follow a normal distribution. Handmade paper is generally not uniform and if you are not an experienced paper maker you will likely have clumps in the pulp. There will be thin spots and thick spots in your paper. Table 6.1 shows actual experimental data for two different sheets of paper. The sample consists of 30 data points taken at random locations on each sheet of paper. The data points have been arranged in numerical order for further analysis.

Now we can compare these two sheets of paper analytically. The first thing we notice is that, on average, sheet No. 2 is thicker than sheet No. 1. We can also extract some information by comparing the standard deviations. For sheet No. 1 we find the standard deviation is $\sigma_1 = 8.7 \times 10^{-2}$ mm and for sheet No. 2 we have $\sigma_2 = 4.7 \times 10^{-2}$ mm. The smaller deviation for sheet No. 2 suggests that it is more uniform or smoother. The ratio of the two standard deviations is about one half but that comparison is a little misleading. If we were comparing sheets of paper that were approximately the same thickness we can compare their standard deviations directly. What if one sheet of paper is thick and the other sheet much thinner? In this case we would be better off comparing

Table 6.1. Thickness at random locations on two different sheets of handmade paper.

Sheet No. 1 Thickness (mm)	Sheet No. 1 Thickness (mm)	Sheet No. 2 Thickness (mm)	Sheet No. 2 Thickness (mm)
0.31	0.57	0.81	0.94
0.38	0.58	0.85	0.94
0.38	0.58	0.87	0.94
0.40	0.58	0.88	0.95
0.43	0.58	0.89	0.95
0.46	0.58	0.89	0.95
0.47	0.58	0.90	0.95
0.48	0.59	0.90	0.95
0.50	0.61	0.90	0.95
0.51	0.61	0.92	0.97
0.51	0.62	0.92	0.97
0.52	0.65	0.93	0.98
0.53	0.65	0.93	0.99
0.54		0.93	1.02
0.55		0.93	1.03
0.55			
0.55			
Mean	Standard deviation	Mean	Standard deviation
0.53 mm	8.7×10^{-2} mm	0.93 mm	4.7×10^{-2} mm
Coefficient of variation		Coefficient of variation	
16%		5.1%	

the ratio of the standard deviation to the mean value. This ratio is called the coefficient of variation (C_v). The coefficient of variation can be calculated as a percentage by the following equation:

$$C_v = 100\% \frac{\sigma}{\overline{x}},$$

where σ is the standard deviation of the thickness and $\overline{x}$ is the mean thickness.

For example, if a thin sheet of paper has a mean thickness of 0.05 mm and the standard deviation is 0.01 mm, then the coefficient of variation is 20%. However, if a thick sheet of paper with an average thickness of 0.25 mm has the same standard deviation of 0.01 mm, the coefficient of variation is only 4.0%. For the sample experimental data we see that sheet No. 1 has $C_{v1} = 16\%$ whereas sheet No. 2 only has $C_{v2} = 5.1\%$. These ratios differ by more than a factor of three.

When we compare measurements we need to have a standard. For a smooth surface the standard deviation or coefficient of variation should be small compared to that of a rough surface. This is easy to say, but just what is meant by small? The best way to determine how much variation there is in a smooth piece of paper is to measure something that was made in a commercial paper mill. In order to demonstrate this, you should measure the thickness, standard deviation and coefficient of variation for a sheet of factory-made paper. Use a piece of printer paper, an index card, or something you regard as a smooth sample. Record the data in the data table marked **Paper-making data sheet—thickness variation of commercial paper**.

We can take the analysis of our paper one step further by plotting the thickness distribution. Calculating the probability distribution function is rather difficult, but a simple way to display the distribution is to make a histogram. The horizontal axis of a histogram contains bins and the vertical axis tells us the frequency or number of times a measurement falls within the bin. The data in table 6.2 was arranged in numerical order so as to make it easy to see how many times a particular value was

Table 6.2. Thickness measurements and corresponding frequency data.

Sheet No. 1		Sheet No. 2	
Thickness (mm)	Frequency	Thickness (mm)	Frequency
0.31	1	0.81	1
0.38	2	0.85	1
0.4	1	0.87	1
0.43	1	0.88	1
0.46	1	0.89	2
0.47	1	0.9	3
0.48	1	0.92	2
0.5	1	0.93	4
0.51	2	0.94	3
0.52	1	0.95	6
0.53	1	0.97	2
0.54	1	0.98	1
0.55	3	0.99	1
0.57	1	1.02	1
0.58	6	1.03	1
0.59	1		
0.61	2		
0.62	1		
0.65	2		
Number of points	30	Number of points	30

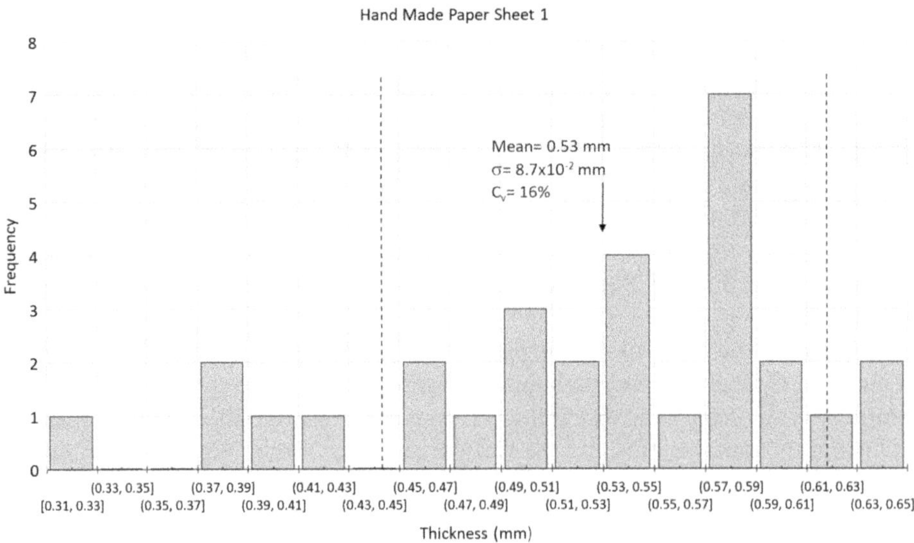

Figure 6.17. Histogram of thickness data for sheet No. 1.

repeated. For sheet No. 1 we measured a thickness of 0.38 mm twice and 0.58 mm six times, as shown in table 6.2.

Now we can compare the two sheets by creating histograms that show the thickness distributions. As we expected, the distribution shown in figure 6.17 looks nothing like a bell curve. The mean thickness is shown by the arrow and the width of the standard deviation is approximately indicated by the position of the dashed line.

Since these distributions are not bell curves or normal distibutions, we cannot apply the 68% rule. Of the 30 data points for each sheet we find that 22 points or about 73% of all the data falls within one standard deviation on either side of the mean.

Points to consider in analysis

Now that you have collected thickness data, you need to interpret the results. These results should be documented in your laboratory report. Remember our overall goal was to see if the force applied while drying the paper influenced the mean thickness and smoothness of the paper. There are several way to accomplish this. The simplest way is to compare the mean thickness, standard deviation and coefficient of variation in each section. Some degree of skill is required to make uniform paper. You may find that the variation in thickness for each section is too large to make a meaningful comparison of each section. There should be a clear distinction between the minimal force and maximum force regions. We can also plot a histogram representing the distribution of thickness values for each section. Figures 6.17 and 6.18 represent two different sheets and we can compare both the mean thickness and the coefficient of variation. Sheet No. 1 has a mean thickness of 0.53 mm and sheet No. 2 has a mean thickness of 0.93 mm so on average it is thicker. How about the smoothness of sheet No. 2? Figure 6.18 shows the coefficient of variation is 5.1% for sheet 2 compared to the much larger variation of 16% for sheet No. 1. So sheet No. 2 is both thicker and smoother. You should conduct a similar analysis for the three different sections of your own paper. To construct a histogram you will need to make a table similar to table 6.2 that shows the frequency or number of times that a particular thickness is measured. Notice that the sample histogram shown in figures 6.17 and 6.18 contains bins for the thickness values, which are plotted along the horizontal axis. You need to determine the width of the bins based on the range of your thickness data. The width of each bin should be the same so as to avoid confusion.

Alternative methods of analysis

Your laboratory instruction may direct you to consider other methods of analysis. One such method is to create a thickness profile of the paper. You can measure the position of each thickness value. This requires determining the horizontal or both the horizontal and vertical coordinates of each data point. From this data a graph of thickness as a function of position can be created. Mark the boundaries of the three different force sections on the graph and see if there is any discernable pattern.

Paper mass can also serve as a diagnostic quantity. For these experiments you use the same volume of pulp each time and wring out the water from the sponge into the pulp vat. Make a series of sheets using the same volume of pulp each time. As you withdraw fiber from the pulp vat and return water from the sponge, the density of the pulp varies. Subsequent sheets are made with pulp that has a higher water-to-fiber ratio. This more watery pulp will make a thinner sheet of paper as long as the same volume of pulp is used each time. You must also keep the area of the sheet

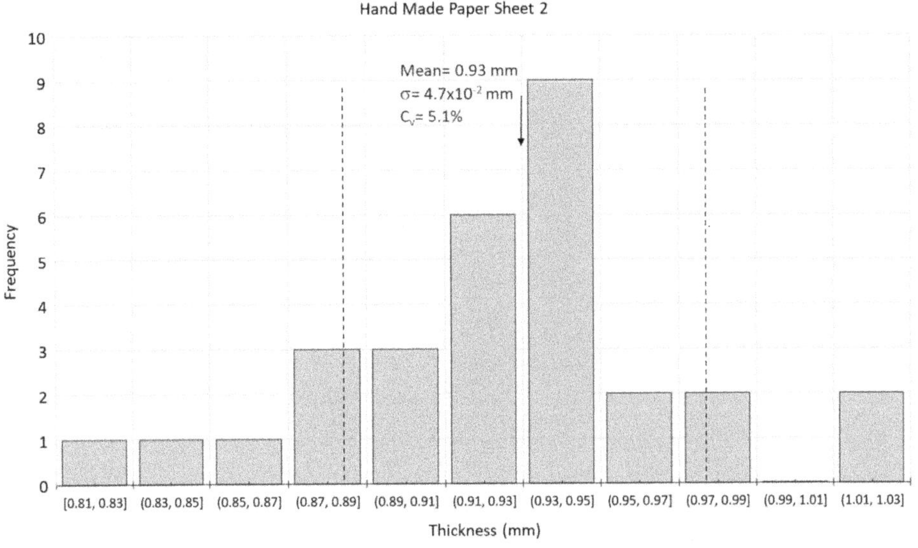

Figure 6.18. Histogram of thickness data for sheet No. 2.

constant by completely filling the same size deckle each time. The process of making a sheet of paper is a little different since you are now pouring a sheet instead of dipping the frame into the pulp vat. As you pour the pulp into the deckle and frame, you must agitate the pulp and try to spread it uniformly. Pouring can be done over the pulp vat or in a low walled tray such as a cookie sheet. Once you have made a series of sheets you can weigh them. Then plot the sheet mass as a function of sheet number. We would expect there to be a progression in mass. The first sheet should have the greatest mass and the last sheet the lowest mass. The mass method is particularly useful when calipers or dial indicators are not readily available. There are, no doubt, many other variations that can be useful for this experiment. Remember that we are using paper making as a tool to introduce the statistical treatment of data and the use of graphs for displaying relationships between experimental parameters.

Data sheets: For a paper-making experiment

Paper Making Data Sheet-Force Variation Page 1 Date____________

Group Names____

Sheet No.__________ Dial scale 1div. = __________mm

Pulp mass (g)__________ Volume (ml)__________ Density (g/ml)__________

Question for Lab report:

Are there any patterns? Does the drying force influence the thickness and/or the smoothness?

Minimal Force		Medium Force		High Force	
Dial Reading	Thickness (mm)	Dial Reading	Thickness (mm)	Dial Reading	Thickness (mm)

Paper Making Data Sheet-Force Variation Page 2 Date____________

Group Names____

Minimal Force		Medium Force		High Force	
Dial Reading	Thickness (mm)	Dial Reading	Thickness (mm)	Dial Reading	Thickness (mm)

Average ________ ________ ________

STDEV ________ ________ ________

C_v % ________ ________ ________

Paper Making Data Sheet- Thickness variation of commercial paper Date______________

Group Names____

Commercial Paper Type:	
Dial Reading	Thickness (mm)

Average _________ STDEV _________ C_v % _________

Question for lab report:

You would expect the commercial paper to be much more uniform than your handmade paper, is that the case? Compare the variations and explain in detail.

6.2 Stress–strain curves of silk thread (丝线应力应变曲线 Sīxiàn yìnglì-yìngbiàn qūxiàn)

Objective

In this experiment we are going to learn about silk and how to harvest silk fibers from a silkworm cocoon. Silk is a remarkably strong substance and we will also measure just how strong it is by measuring its breaking strength and elastic modulus.

Chinese vocabulary

丝 sī Silk, thread or a thread-like thin, filament	应力 yīng lì Stress
丝线 sī xiàn Silk thread	应变 yìng biàn Strain
蚕 cán Silkworm	曲线 qū xiàn Curve
蚕茧 cánjiǎn Silkworm cocoon	杨氏模量 yángshìmúliàng Young's modulus
精练 jīngliàn To degum silk	缫丝 sāo sī Reeling silk from a cocoon
养蚕 yǎngcán Sericulture or the raising of silkworms	

Materials

For silk reeling	
Silkworm cocoons	Hotplate or hot water source
Washing soda (sodium carbonate)	Hair conditioner
Safety glasses	Sewing bobbins
Beaker (500 ml)	Nut and bolt to fit sewing bobbin
Measuring spoon	Blow dryer
For stress–strain measurements	Camera suitable for low light
Electronic mass scale	Nail polish with a bright color
Low friction pulley	Laboratory balance weight 100–200 g
Vernier caliper	Meter stick
Ring stand to hold pulley	Three-finger clamp
Microscope with stage micrometer or laser pointer or known wavelength	

Introduction

Like most ancient discoveries, it is not entirely clear just what happened but it seems that silk was discovered by accident. As the story goes, the Empress 嫘祖 Léi Zǔ was having tea in the shade of a Mulberry tree when something fell from the tree. A silk cocoon (蚕茧 cánjiǎn) fell right into her tea cup. She did not throw the tea out but instead observed the cocoon. After a while, she noticed that it changed its shape and left behind a mat of fibers. The fibers were like a spider's web but not as sticky. From this simple observation she invented the process of reeling the fibers to make thread.

In recognition of her discovery, 嫘祖 Léi Zǔ is often regarded as the goddess of sericulture (养蚕 yǎngcán). We are not certain about just when silk production began in China but fragments of silk have been found dating back to Neolithic times (Kuhn 1984). The oldest known fragments from the domesticated silkworm *Bombyx mori*, were found at an archeological site in a layer dated to approximately 2750 BC (Barnard 1975). By the 14th century BC, during the Shang Dynasty (商朝 Shāng cháo), the sericulture industry was well established and China began exporting this unique material. The famous Silk Road was established and silk became a major export. Europeans had a great desire for the new material and had no idea how it was made. The actual method of raising silkworms and the production of silk was a state secret.

Bombyx mori silkworms extrude two important proteins that form the silk fiber. Fibroin forms the main component of the two filaments that make up a single strand of silk. These two filaments are held together with a gum-like material composed of the protein sericin. This sericin gum layer can be dissolved by a warm alkaline solution as we will see in this experiment. Figure 6.19 shows an electron micrograph of a silk strand. The sericin gum layer has not been completely dissolved and can be seen partially covering over the fibroin. In the upper right portion of the image we see what looks like a seam, indicating that the strand has two component parts known as brins.

The scale bar in figure 6.19 is 120 μm long and the width of the silk strand is about 10 μm. To put this in perspective we can compare the strand to the size to a thick human hair, which is about 100 μm in diameter. From the micrograph it is obvious that the strand is not a cylinder, but for simplicity we will assume that it has a circular cross-sectional area.

In the first part of this experiment we will learn how to harvest the silk fibers from a cocoon (蚕茧 cánjiǎn). This process is known as silk reeling (缫丝 sāosī). The second part of the experiment will involve testing the elasticity of our threads. Cocoons are hollow and there may be a dead silkworm in each cocoon. Some people work with 'Peace Silk' cocoons. These cocoons have a hole at one end where the pupa made its way out. The cocoon is spun from a single strand that is on the order of 1000 m in length. If the pupa emerges from the cocoon, the continuity of the strand is destroyed and one is left with many shorter fibers. Of course, for breeding purposes, some pupa are allowed to emerge from the cocoon as a moth.

Figure 6.19. Electron micrograph of a silk strand.

Experimental

The first step is to degum the cocoon (精练 jīngliàn) so we can access the fibers. We will use washing soda and hot water to remove the gum. Washing soda is not ordinary laundry detergent or powdered soap. It is a strong alkaline chemical also known as sodium carbonate (Na_2CO_3). This is not something you want to get in your eyes so you must wear safety glasses. The sericin coating will easily dissolve and will produce a mat of fibers. If there was a silkworm in the cocoon it will now be visible. We would like to determine how much silk can be obtained from a single cocoon. Be sure to weigh the dry cocoon, silkworm, debris in cocoon and the dry silk mat.

Procedure to degum the cocoon

(1) Add about 1 or 2 teaspoons of washing soda to a 3/4 full beaker (375 ml) and heat the water. Do not boil the water.
(2) Place the cocoon into the hot water and stir with a spoon. The gum will dissolve and soon you will see the cocoon fall apart. The worm will be visible.
(3) Use a spoon to retrieve the worm and the mat of silk fibers.
(4) Dry the fiber with a blow dryer and weigh the dry fibers.
(5) Weigh the worm.

Using the weight of the dry silk fibers calculate the number of cocoons required to make a silk dress of mass 300 g. Comment on this in your laboratory report.

Silk reeling

For these experiments we wish to slowly dissolve the gum. It is difficult to reel silk from a totally degummed cocoon since it simply falls apart and you get a mess. You can reel the silk onto an ordinary sewing machine bobbin (figure 6.20). Prepare

Figure 6.20. Bobbin mounted on a machine screw. For long lengths of silk thread, insert the bobbin into a cordless drill.

several bobbins as shown below. You need to insert a long machine screw into the bobbin to act as a handle. Secure the screw with a nut. If you plan to reel many meters of silk, you may wish to use a cordless drill to slowly spin the bobbin. We will make thread consisting of a single strand, three strands and five strands.

(1) Use the toothbrush to clean 'dead ends' off of the cocoon while it is still dry.
(2) Start with fresh water and soak the cocoon in cold water for a few minutes.
(3) Remove the cocoons from the water and heat it up. Do not boil the water.
(4) As the water is heating add conditioner, a fraction of a spoonful of washing soda and a few cocoons. You will need at least five cocoons. Do not boil the water.
(5) Once the water is hot, immerse the cocoons under water and they will fizz. This is air escaping from the cocoon. Once they stop fizzing the cocoons will not be as buoyant but they will still float.
(6) Move the toothbrush over the surface of the water to catch a fiber. Once you have a fiber, pull on it and try to get a strand of a few meters in length. If the strand breaks you might have snapped it or it just might be a short broken fiber near the surface. At some point you will find that you can pull and pull a seemingly endless fiber. That is likely to be the main fiber. Drape the fiber over the edge of the beaker.
(7) Now repeat this procedure for each cocoon keeping the individual fibers separated from each other.

Drape all the fibers over the edge of the beaker, as shown in figure 6.21.

Now you are ready to make thread. One lab partner should help monitor the silk as it comes off the cocoons and keep the strands from becoming tangled. The second partner can twist the thread and the third can reel it onto a bobbin. You should try to dry the silk with paper before it is wound onto the bobbin.

(1) Use your thumb and forefinger to twist the fiber so that the twist propagates up to the edge of the beaker.
(2) Attach this thread to a bobbin and slowly rotate the bobbin.
(3) Repeat this procedure until you have reeled several meters of thread around the bobbin.

Figure 6.21. Cocoons floating in warm water with silk strands ready for reeling.

You will need to make several bobbins. One should just have a single fiber. Also make bobbins with three and five fibers per thread. Label each bobbin with some tape. In the second part of this experiment we will test your silk thread under stress.

Part 2: Stress–strain testing

In our previous experiment we learned the fine art of silk reeling. In this experiment we are going to subject our silk thread (丝线 sī xiàn) to an applied force and then measure how much it stretches. The applied force will be directed along the long direction of the thread. When subjected to this type of load we say that the thread experiences tensile stress. Tensile means that we are stretching and not compressing the strand. The physics definition of stress (应力 yīng lì) relates the applied force and the cross-sectional area of the strand. The force F is applied to the strand as shown in figure 6.22. For simplicity we will assume the strand is cylindrical in shape. The cross-sectional area of the fiber is simply the area of a circle ($A = \pi r^2$). To calculate the area we will need to measure the diameter of the thread. Several methods for measuring the diameter are discussed below.

Stress tells us how the applied force is distributed over the cross-sectional area of the strand and is given by

$$\sigma = F/A.$$

The units of stress are N m^{-2}, which is also known as the Pa (Pascal). The cross-sectional area of an individual silk fiber is very small so even a few grams of applied mass can cause very large stresses. Large stresses are measured in units of GPa (10^9 Pa) and MPa (10^6 Pa). Suppose we suspended a 1.00 g mass from the end of a strand that has a diameter of 100 μm. The weight of a 1.00 g mass corresponds to a force of

$$F = mg = 1.00 \times 10^{-3}\ \text{kg} \times 9.80\ \text{m s}^{-2} = 9.80 \times 10^{-3}\ \text{N}.$$

Figure 6.22. A cylinder subject to tensile stress.

Since the strand is a cylinder with a diameter of 1.00×10^2 μm (1.00×10^{-4} m), we can calculate the area as follows. The radius of the strand is 50.0 μm or 5.00×10^{-5} m.

$$A = \pi r^2 = \pi \times (5.00 \times 10^{-5}\,\text{m})^2 = 7.85 \times 10^{-9}\,\text{m}^2.$$

The stress would then be

$$\sigma = F/A.$$

or $\sigma = 9.80 \times 10^{-3}$ N/7.85×10^{-9} m^2 = 1.25×10^6 Pa or σ = 1.25 MPa. Keep in mind that 100 μm is roughly the diameter of a thick human hair and larger than a single strand of silk.

A second important parameter is called strain (应变 yìngbiàn). When we apply tensile stress to the strand it will stretch. We quantify the amount of stretch by measuring the change in length (ΔL) and dividing that by the original length (L_0). Strain is often expressed as a percentage, but not always. We can write the strain as

$$\varepsilon = \Delta L/L_0.$$

Our silk fibers can be stretched up to a few tens of percent without breaking. If the original length of the fiber was 2.00×10^3 mm long and we stretched it by 20.0 mm, the strain would be

$$\varepsilon = 20.0\,\text{mm}/2.00 \times 10^3\,\text{mm} = 1.00 \times 10^{-2} = 1.00\%.$$

Silk behaves as an elastoplastic material. Figure 6.23 shows a typical stress–strain curve for a single silk strand. There is a well-defined linear region up to about 0.9% strain. We can judge the linearity of the stress–strain relation by calculating the R^2 value of a linear fit to the data. For this data, we find that $R^2 = 0.996$, indicating a strong linear relationship between stress and strain. At higher strains the curve bends over and is clearly non-linear. This non-linearity is an indication of strain-hardening or plastic behavior.

The linear relationship between stress and strain means that an increase in the strain produces a proportional increase in stress. We can express this mathematically as

$$\sigma = Y\varepsilon.$$

Since stress (σ) is on the vertical axis and strain (ε) is on the horizontal axis, Y represents the slope of the elastic region. This slope is called the elastic modulus or Young's modulus (杨氏模量 yángshìmúliàng).

Strain is a ratio of two values with the same units, so the units cancel out. This means that the units of Y are the same as those of stress (GPa or MPa). We all know from experience that if you pull on a string with enough force, the string will break. Before it breaks it may undergo a permanent deformation. That means if you release the force, the string will not go back to its original length. It may be a little longer since it is now deformed. Our silk fibers will behave elastically for low force loads and then they start to deform. Eventually the force will be great enough to break the fiber.

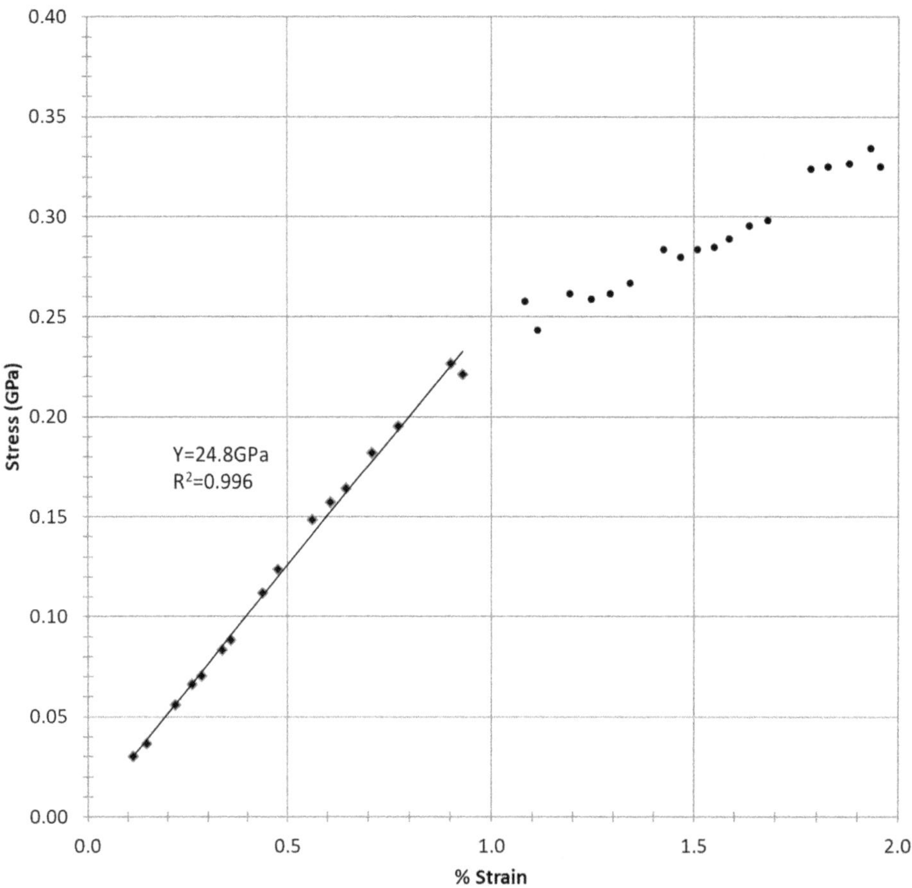

Figure 6.23. Stress–strain curve for a single silk strand.

Stress–strain measurements

The basic stress–strain apparatus is set up as shown in figure 6.24. We can determine the tension in the fiber (F) by measuring the change in the apparent weight of the 100 g mass on the scale. The fiber is attached to the mass and the caliper stem with nail polish. Nail polish makes a good glue and is easily removed with acetone. To stretch the fiber simply press on the caliper and close the jaws. Before you take data, put a little tension in the fiber so that it does not sag. Once you have removed the slack you can zero the mass scale and the caliper. Gently stretch the fiber and record the change in weight and the displacement (stretch) in a data table. For each mass value calculate the force F applied to the strand. From the change in length calculate the strain ε. Repeat your measurements several times. Do not stretch the strand to its breaking point until the last run. Record the breaking force and the corresponding breaking strain. If the silk strand is too long you may have to strain it beyond the travel of the caliper. In this case you can simply pull on the strand with your finger and record the effective mass at which it breaks. A very effective means of doing this is to use a phone

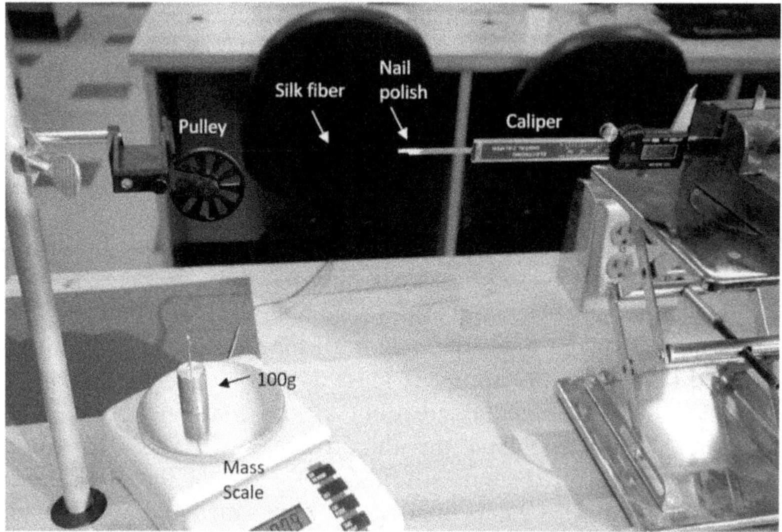

Figure 6.24. Stress–strain apparatus.

or similar device to record a video of the mass reading. Slowly pull on the strand and then say something to indicate that it snapped. You can then review the video to determine the exact breaking force. Determine the breaking force and strain for a single strand and the threads you made from three strands and five strands.

Procedure for stress–strain analysis

(1) Assemble the stress–strain apparatus as shown above in figure 6.24. Be sure to measure the starting length of the thread **before** you make any stress–strain measurements. Calculate the change in length that corresponds to a strain of 2% and make a note of it on your data table.
(2) Attach one of your silk threads to a laboratory weight of known mass. Several hundred grams should be sufficient. You will need a thread length of about 1 m. Since the silk is difficult to see, use brightly colored nail polish to mark the ends of the threads. Nail polish makes a good glue and is easily removed from the apparatus at the conclusion of the experiment.
(3) Pass the free end of the thread over the pulley and attach it to the stem of a caliper. The caliper should be extended at this point and the thread should have some slack. If there is too much slack to start with, you may not be able to utilize the full travel of the caliper.
(4) Record the apparent mass with the silk thread still slack.
(5) Be sure that the caliper stem is in line with the pulley, so that the stress is applied along the silk thread without any sideways forces.
(6) Grasp the caliper jaws and gently begin to close the gap. The silk thread should tighten up and the apparent mass should decrease. Record the apparent mass and displacement in your data table. You can carefully zero the caliper when you just detect a small change in the apparent mass.

(7) Continue to close the gap while recording the force and change in length. You may wish to take data at a fixed mass change interval of about 0.1 g. You have calculated the 2% strain length and you can use this to judge when you are in the plastic range.

(8) Once you have achieved a 2% strain, plot your data to see if there is a linear region. You should continue to strain the thread until you have made the transition into the plastic (non-linear) range.

(9) Since a single strand is capable of tens of percent strain you may not have enough caliper travel to break the fiber.

(10) If the caliper has reached its full range with the thread intact, you can use your finger to continue stretching until the strand breaks. Use a phone to record a video of the apparent mass as you stretch the thread. This will enable you catch the last reading at the snapping point. Record the apparent mass at which the thread breaks. Be sure to compare the breaking force for a single strand and threads of three and five strands. Does a thread of 5 strands break at five times the breaking force of a single strand? Comment on this in your lab report.

(11) Plot your force vs. strain data. Compare the slopes for a single strand and threads of multiple strands. To convert the force to stress, you will need to determine the diameter of the strand. Depending on the objectives of your experiment you may not need to make these additional measurements. Check with your laboratory instructor. Please see the section **Methods for measuring the diameter of the silk strand** for additional information concerning these measurements.

Analysis

Is the thread made of three fibers three times stronger than a single fiber? Try to experimentally determine this. Exactly what is meant by 'stronger' is something you can choose to quantify. For example, one material may be described as stiffer than another material. This means that for the stiffer material a greater force is required to achieve the same strain. If a greater force is required to produce same amount of strain, then the slope of the stress–strain curve would be greater. This is reflected as an increase in the Young's modulus. In engineering terminology, stiffness is not the same thing as strength. In fact, there are quite a few different ways to quantify the strength of a material and each one has a precise engineering definition. Strength can also be quantified by the maximum amount of stress a material can withstand while remaining in the linear or reversible portion of the stress–strain curve. The non-linear portion of the stress–strain curve is often referred to as the plastic region. Plastic, in the engineering sense of the term, means unrecoverable strain. When the stress is decreased the strain will not decrease in the same proportion and the material will have undergone a non-reversible change in shape. Elastic materials such as silk, rubber and spider silk can undergo large amounts of plastic deformation before finally breaking. The stress at which the plastic deformation begins is called the yield strength or yield stress. Another measure of the strength of a material is the ultimate strength. This is the maximum stress that a material can withstand before breaking. In comparing your silk threads you may wish to compare any or all of these measures of strength. There is a great deal of variability in the mechanical properties of silk. Even a single strand of silk sampled in several places will not have exactly the same mechanical properties. Zhao *et al* (2007) explored the variability in Young's modulus for silk strands of different diameters. They report values of approximately $Y = 20$–30 GPa for strands of 12 μm diameter. The slope of the linear region shown in figure 6.23 gives a Young's modulus of $Y = 24.8$ GPa. Our experimental results show a departure from linearity at a yield stress of approximately 0.20–0.25 GPa, which is also consistent with Zhao's values for a 12 μm diameter strand. In terms of yield stress this is comparable to ASTM-A36 structural steel, which has a minimum yield strength of 0.25 GPa. Of course, A36 steel is much stiffer, with a Young's modulus of about 200 GPa (ASTM 2014).

Methods for measuring the diameter of the silk strand

Photographic technique

In our calculation of stress we need to know the cross-sectional area of the strand. For simplicity, we can make a graph of force as a function of strain. Such a graph will show the same shape, since the stress is just force divided by a constant value of the cross-section area. If we want to compare silk to other engineering materials, we do need to convert the force into stress. Since the silk strands are on the order of 10 μm, how can we possibly measure them? The simplest techniques require the use of photography and scaling factors. A microscope and some type of size scale such as a hemocytometer or stage micrometer can be used. The picture shown below in figure 6.25 is of a single silk fiber on a hemocytometer. You can carefully hold a

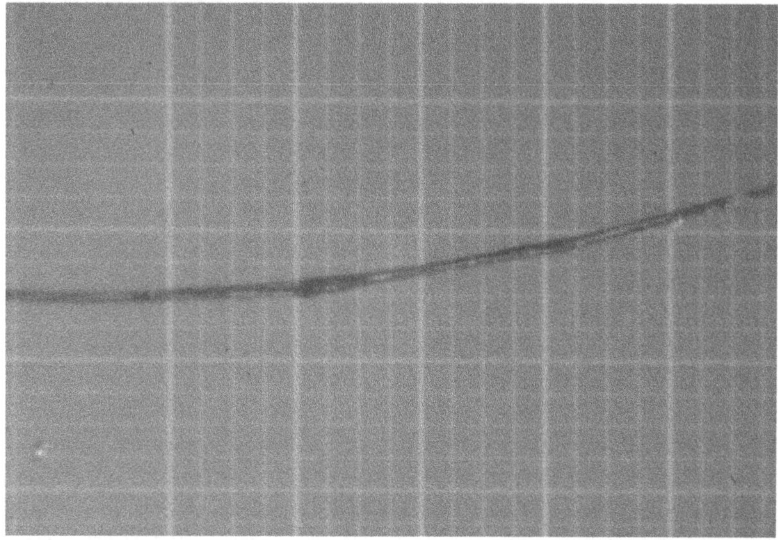

Figure 6.25. Single silk strand photographed with a hemocytometer as a size scale.

camera to the eyepiece of a microscope and take a similar photo. Normally a hemocytometer is used for counting blood cells but it also makes a great size scale marker. The small squares are 50 μm on edge. We can compare the diameter of our strand to the 50 × 50 μm scale. You will notice that the fiber is not uniform in diameter. Some of the irregular shape is caused by gum remaining on the strand and there is also a natural variation in strand diameter. We should use an average diameter for our calculations.

Most photo editing software, such as Photoshop®, has the capability of displaying the cursor position (figure 6.26). It is important to make such measurements in units of pixels. Pixel coordinates are not altered by changes in the magnification of the image.

We can calculate the diameter of the silk strand by first determining the number of pixels there are in a known distance. In this image there are 22 px in a distance of 50 μm. These values are given to two significant figures. If a feature has a size of 10 px on the photograph it has an actual size of

$$10\text{ px}\left(\frac{50\ \mu\text{m}}{22\text{ px}}\right) = 23\ \mu\text{m}.$$

Diffraction technique

Another method of measuring the diameter of the silk strand is to use diffraction. When light passes through an aperture whose size is comparable to the wavelength of the incident light, there is a noticeable deflection of its path. Visible light has a wavelength of several hundred nanometers or a few tenths of a micrometer. Thus, a slit or aperture the size of a silk strand will produce a very significant diffraction pattern. There is a

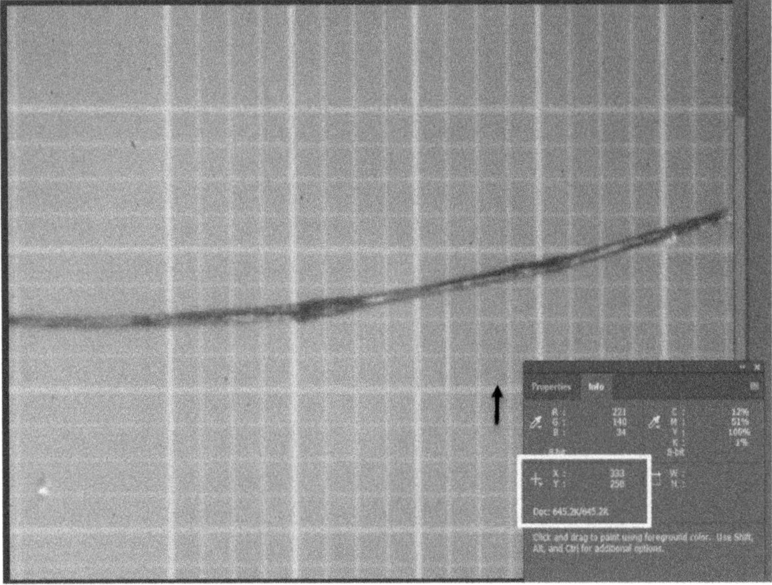

Figure 6.26. The info window in Photoshop® showing the coordinates of the cursor in units of pixels.

concept in optics known as Babinet's principle, which tells that a narrow solid object will produce a diffraction pattern similar to a slit of the same size. Based on this principle, we can treat the diffraction pattern of a single silk strand as if it is produced by a slit of similar diameter. This makes it rather simple to measure the diameter of the strand by using a common laser pointer. Figure 6.27 illustrates this idea.

The laser pointer is held in place with a three-finger clamp. Brightly colored nail polish is used to adhere the silk strand to a metal frame. Some nail polish is applied to the fiber to make it easier to handle.

To show how such measurements can be used to determine the strand diameter, a blue laser pointer with wavelength $\lambda = 4.05 \times 10^2$ nm is used to illuminate a fine wire. This produces a classic diffraction pattern as shown below in figure 6.29. The wire used is No. 36 AWG, which has a diameter of 1.27×10^{-4} m or 127 μm. That makes the diameter about ten times the diameter of a typical silk strand. The lower portion in the figure shows a measuring tape positioned just above the diffraction pattern. We can use the tape to enable us to find the angular positions of the intensity minima. The intensity of the diffraction pattern varies and there are places where the intensity drops to nearly zero. We can characterize the diffraction pattern by measuring the angular position of these minima with respect to the central bright spot. In order to determine the angular position of a minimum, we need to know the distance from the diffracting object to the screen (d) on which the diffraction pattern is projected. The geometry of the setup is shown schematically in figure 6.28.

The angular position of the first diffraction minimum (θ_{min}) is related to the diameter (a) of the wire or silk strand by

Figure 6.27. Silk strand illuminated by a laser pointer for diffraction experiments.

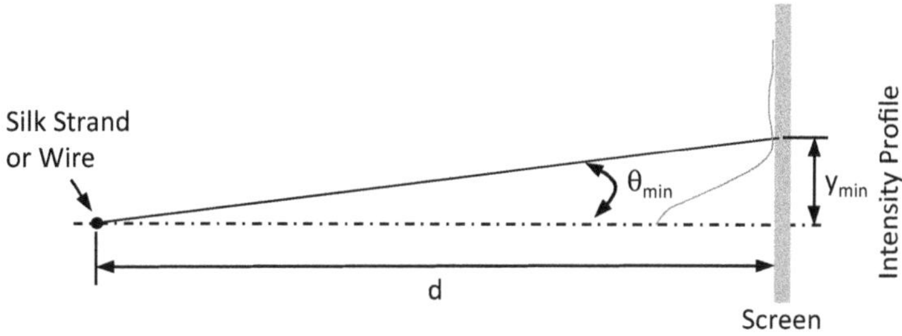

Figure 6.28. Geometry of the diffraction measurement.

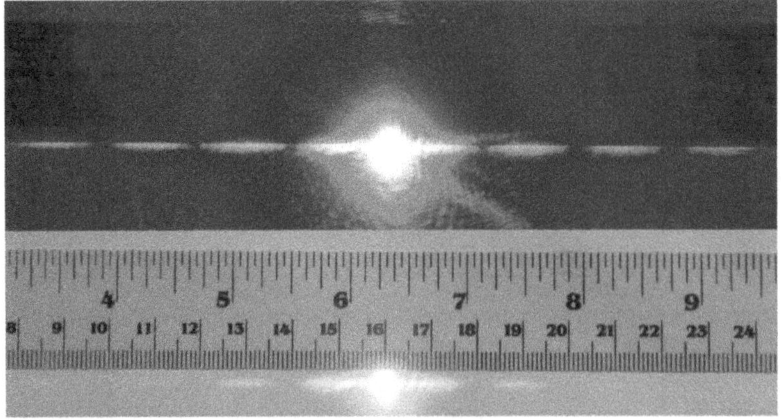

Figure 6.29. Diffraction pattern of a fine wire (No. 36 AWG) using a laser pointer of wavelength 405 nm. Top image taken in a dark room, bottom image with lights on showing the measuring tape.

$$\lambda = a\sin(\theta_{\min}).$$

From the geometry shown in figure 6.28, we can relate the distance to the screen (d) to the position of the minimum ($y_{\min}$). This is given by

$$\tan(\theta_{\min}) = \frac{y_{\min}}{d}.$$

For angles less than about 15° we can approximate the tangent of the angle with the sine of the angle. The angle to the first minimum is about 0.18° so this approximation is valid.

$$\tan(\theta_{\min}) = \frac{y_{\min}}{d} \cong \sin(\theta_{\min}).$$

Using this approximation, we can relate the diameter (a) to the position of the first minimum ($y_{\min}$) by

$$a = \frac{d\lambda}{y_{\min}}.$$

From the diffraction pattern for the wire shown in figure 6.29 we can see that the position of the first minimum is about 2.10 cm. For these measurements, the distance to the screen is 6.70 m and the wavelength is 405 nm, which yields a diameter of 129 μm. This is only about 1.6% larger than the standard wire gauge diameter.

Similar measurements can be made for the silk strand. As figure 6.30 shows, the diffracted light is spread over a much wider distance.

Black tape with a hole in it is placed at the central bright spot. This decreases the intensity of the reflection and improves the overall contrast of the photograph. The intensity of the side maxima are rather faint and do not show up with the room lights on. To determine the distance to the first minimum, we can use a photographic technique similar to that described for determining the diameter with a microscope. The diffraction pattern is photographed with the lights on and off while keeping the magnification of the image constant. With the lights on, we take a photograph of the measuring tape and use

Figure 6.30. Diffraction pattern of a single silk strand using a laser pointer of wavelength 405 nm. Only one side of the pattern is shown.

the cursor technique to determine the scale factor in pixels per centimeter. Then, we photograph the diffraction pattern with the lights off and the camera in the same position. Of course an even simpler method is to simply use a pen to mark the position of the first minimum with the lights off and then measure the distance to the central spot with the lights on. The method you use depends on the objectives of the laboratory instructor. For the image of the silk strand shown above in figure 6.30, we found the scale factor to be 2.00 cm/64 px and the distance from the central spot to the first minimum to be 358 px. This yields a position of about 10.88 cm from the center with the screen at 3.18 m. The angle to the first minimum is then $\theta_{\text{min}} = 1.96°$, which is well within the small angle approximation. Substituting these values into the equation for the diameter yields $a = 11.8$ μm. This result is consistent with our assertion that the diameter is about 10 μm. It is well known that the diameter varies along the length of a silk strand. While our equipment may seem crude compared to electron microscopy, literature values for the diameter of *B. mori* silk have been found to be in the range of 10–16 μm, including the sericin sheath of about 1 μm in thickness (Zhao *et al* 2007).

Data sheets: For silk stress–strain experiments

Data Sheet for Stress-Strain Measurements Date____________

Group Names____

Initial length of thread (L_0) _______________(m) Change in length at 2% strain__________

Mass of laboratory weight with silk attached but still slack________

Number of strands ________

Diameter of thread (μm) __________ Cross-sectional area ___________ (m^2)

Apparent Weight (g)	Weight Change (g)	Applied Force (N)	Caliper Reading (mm)	Length Change ΔL (mm)	Strain ε (%)	Stress σ (GPa)

Data Sheet for Stress-Strain Measurements (continued) Date____________

Group Names____

Diameter of thread (μm) __________ Cross-sectional area ___________ (m^2)

Apparent Weight (g)	Weight Change (g)	Applied Force (N)	Caliper Reading (mm)	Length Change ΔL (mm)	Strain ε (%)	Stress σ (GPa)

6.3 Steelyard balance (秤 Chèng)

Objective

Like most of the experiments in this series there are several different objectives. Our first objective is to build a working 秤 chèng or steelyard balance. Building and understanding the operation of the 秤 chèng will help reinforce ideas about torque and static equilibrium. We will then test the accuracy of our balance and learn to distinguish random error from systematic error.

Chinese vocabulary

杆秤, 秤 Gǎnchèng or chèng Steelyard balance	力 lì force
力矩 lìjǔ, 转矩 zhuànjǔ Torque	力臂 lìbì lever arm
支点 zhīdiǎn fulcrum (for a lever)	重量 zhòngliàng weight unit; weight
重力 zhòng lì weight	两 liǎng unit of weight equal to 50 g
斤 jīn unit of weight equal to 500 g	

Materials

Plastic bowl	Safety glasses
1/2″ Diameter SCH40 PVC pipe (Approx. 80 cm)	Flat washers (#8, #10)
1 1/4″ Diameter SCH40 PVC pipe for counterweight	Steel bailing wire
1 1/4″ Diameter end caps for PVC pipe (2 ea.)	Hacksaw
Sand for counterweight	Needle-nose pliers
3/8-16 Bolt and nut, nut must make tight fit in PVC pipe	Diagonal cutter
Nail and drill bit that will give a snug fit	Sand paper (various grits)
Fishing weights 1 oz, 1/2 oz, split shot 3/10 oz	Files
Assorted laboratory weights	Electronic mass scale
Bolt cutter	Support bars
Cordless drill or drill press	Self-adhesive ruler tape
Flathead screws and corresponding drill bit	Bubble level with screw holes

Introduction

The 秤 chèng that we are going to build is shown below in figure 6.31. No one knows for sure when the 秤 chèng was developed and it is not unique to China. It is known to have been in use during the Han Dynasty (漢朝 Hàncháo), which places it at about 200 BC (Needham *et al* 1962). This type of balance is still in use today and can be found in markets all over Asia. The same basic form can be found in a wide range of sizes from pocket devices to large machines capable of weighing an entire wagon.

In the 16th century, Friar Gasparda Cruz traveled in Asia and wrote one of the earliest European accounts of China. The extensive use of the 秤 chèng was something that caught his attention and led him to comment:

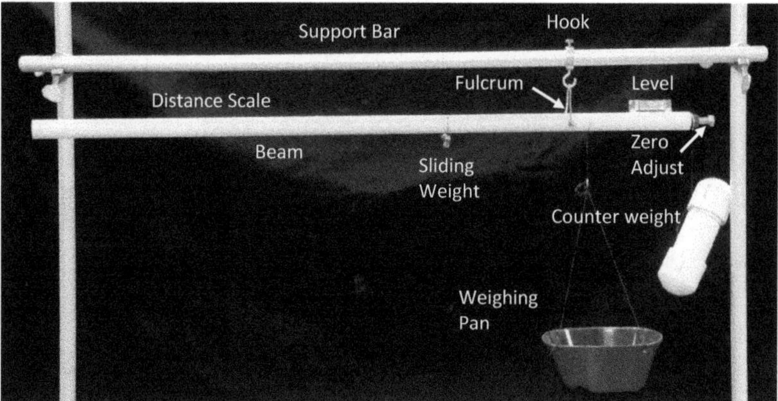

Figure 6.31. Overview of the 秤 chèng or steelyard balance.

> Everyone that goeth to buy in the market, carrieth a weight & balance, & broken silver, and the balance is a little beame of Ivorie with a weight hanging at the one end with a string & on the other end a little scale, and the string of the weight runneth along by the beame, which hath his markes from one Conderin to ten, or of one Maes unto ten (Roth 1912).

The basic principle is rather simple. You place your items in the weighing pan; which for us is a plastic bowl. A sliding weight is then moved along the beam until the beam is level. Then you read the position of the sliding weight. The scales in China often have raised markings that subdivide the length of the beam. We will use a measuring scale that looks more like a ruler, as shown in figure 6.32.

Experimental: construction

The method of using the 秤 chèng will be described a little later. Now it is time to begin construction.

(1) First, cut a section of PVC pipe approximately 80 cm long.
(2) At a distance of about 15 cm from one end of the pipe, drill a hole that is slightly smaller than the diameter of the nail. Insert the nail into the hole to check the fit. The nail should not slide in the pipe. Be sure that the nail hole is straight and perpendicular to the pipe. A drill press is the best method for this. Be sure that the bowl notch is on the **top side** of the beam so that the string hangs below it (figure 6.33).
(3) Make a notch for the weighing bowl. The weighing bowl must be held in a fixed position and the notch keeps the string from sliding. Using a hacksaw make the notch deep enough so that the string will not slip but not so deep as to weaken the pipe. The distance from the notch to the fulcrum **is critical**. Any error here will cause all the mass readings to be off by a constant factor. The measuring tape is calibrated in inches and centimeters. If you intend to use the inch scale, then cut the pan notch at a distance of 1.00 in. from the fulcrum. You may feel more comfortable with

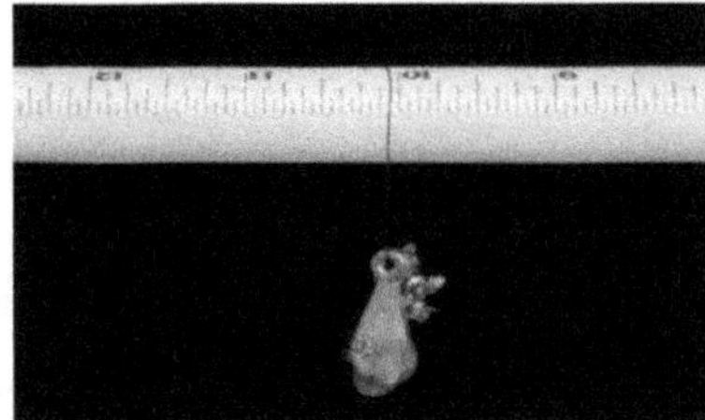

Figure 6.32. Sliding weight and distance scale.

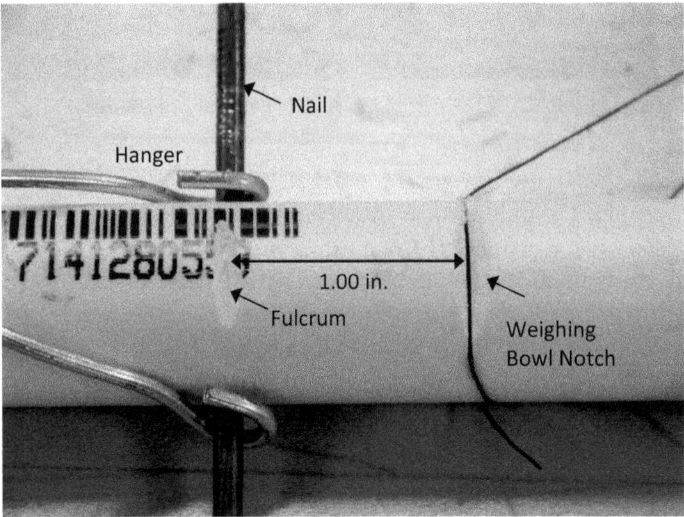

Figure 6.33. Inserting a nail to create the fulcrum and making a notch for the weighing bowl.

using the centimeter scale. In that case you can cut the pan notch at 2.00 cm. The distance you use will show up in the equation for the weight. It is easy to divide by 2 in your head but not so easy to divide by 2.54, which is the conversion factor for centimeters to inches.

(4) Next, we form the hanger from a piece of steel wire. The wire should have two loops to accommodate the nail. The hanger has a 'U' shape with loops on each end (figure 6.34). A bolt cutter can be used to cut off the head and point of the nail. Be sure to **wear safety glasses** when cutting the nail. Also, cup your hand over the part to be cut. This prevents the cut off material from flying out and hitting someone else. Slide one loop over the long part of the nail that sticks out of the pipe and position the second loop over the short part of the nail that sticks out of the opposite side of the pipe. Once the nail is in both loops, press the nail against the table top and insert it deeper into the pipe so that there are approximately equal nail lengths sticking out of the pipe.

(5) Now we will work on the zero adjustment. In order to make a fine adjustment of the zero position we use a bolt that can be screwed in and

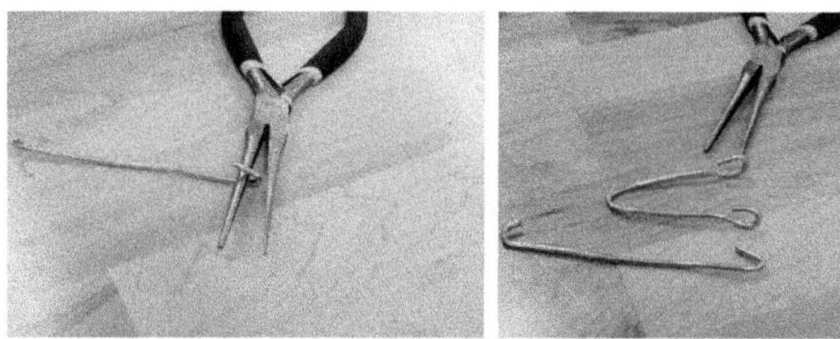

Figure 6.34. Forming the hanger with needle-nose pliers.

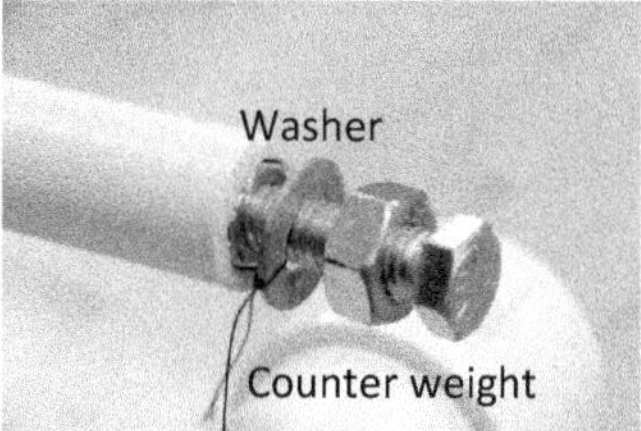

Figure 6.35. Attaching the zero adjustment and counterweight washer.

out of the end of the PVC pipe. The pipe is not threaded so we have to first insert a nut into the end of the pipe (figure 6.35).

The nut must be a very tight fit and it may be necessary to tap it in place with a hammer. Tap the nut gently until it is flush with the end of the pipe. A second nut is used to lock the position of the bolt once the balance is zeroed.

(6) In between the two nuts place a washer. This washer will be used to support the counterweight.

The counterweight washer must always be held securely in place. If it is free to wobble, you will not be able to properly zero the beam.

(7) Now it is time to add a measuring scale to the beam. This is done with adhesive tape that is marked as a ruler. This tape has both centimeter and inch scales. Be sure to line up the end of the ruler tape with the center of the fulcrum. Do not cut the tape directly at the line. Allow a little tape to extend past the start of the scale. This will assist you in aligning the tape with the fulcrum. This step is **also critical** and can create a significant measurement error. Since the tape is marked in one-foot intervals and your beam is longer than one foot, it will be necessary to overlap two sections of tape (figure 6.36).

(8) Next we set up the weighing pan. This consists of a plastic bowl supported by several strings. The strings from the bowl are connected to a flat washer. Another string goes from the flat washer up to the slot on the beam (figure 6.37).

Use the cordless drill to drill small holes in the plastic bowl and attach the support strings first to the bowl and then to the washer. Make sure that the length of each string is adjusted to keep the bowl level. Tie a second

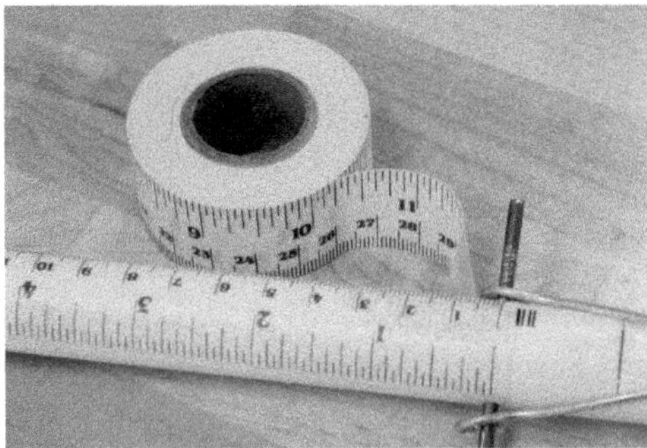

Figure 6.36. Adding the self-adhesive measuring tape.

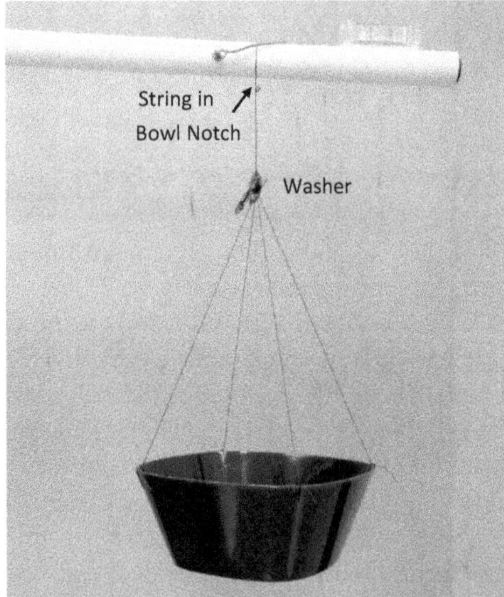

Figure 6.37. Weighing bowl attached to a beam.

string to the washer and attach the other end to the beam at the bowl notch. Tie the string in place so that it does not slip out of the notch.

(9) All of our measurements require the beam to be level. This is accomplished by attaching a bubble level to the beam. The level is held in place by two flathead screws. Drill a hole slightly smaller than the screw and attach the level as shown in figure 6.38.

(10) In this step we will prepare the counterweight. The counterweight balances the weight of the long section of pipe. Cut a 3 in. long section of the 1 1/4 in. diameter pipe. Press an end cap onto the pipe. Tie a support string to the pipe and cover the other end with a cap. You now have an empty counterweight. We will adjust the counterweight by adding sand to it.

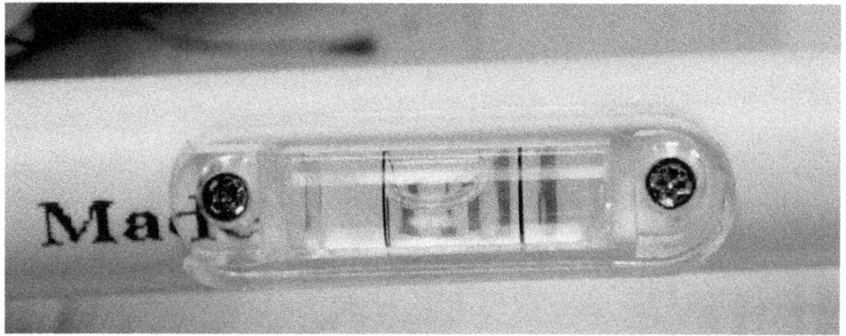

Figure 6.38. Attaching the bubble level with flathead screws.

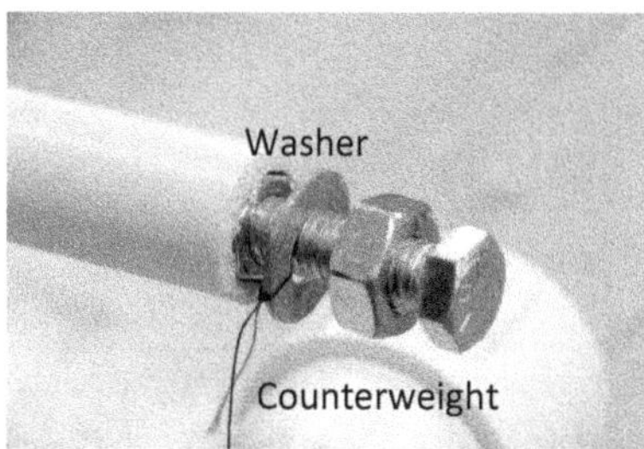

Figure 6.39. Attaching the counterweight to the end of the beam.

(11) Attach the counterweight to the nut at the end of the beam by using a washer, as shown in figure 6.39.

The washer is held against the end of the pipe with the nut that is free to turn, not the one wedged into the end of the pipe.

(12) Use the hanger to attach the beam to the support bar. Adjust the nut to hold the washer in place but leave about half of the length of the bolt sticking out of the end of the pipe. The bolt is our zero adjustment. Proper zero adjustment means that the scale is in perfect balance and the beam is level. To determine the mass of the counterweight attach some calibrated laboratory weights to a string to serve as a substitute for the counterweight. Adjust the weights so that you can achieve balance. Make sure that you still have about half the length of the bolt sticking out. Once you have determined the approximate mass of the counterweight add sand to the counterweight pipe. Be sure that the total mass of the counterweight includes the end caps. Close the end and check to see if the beam is level. Once you have achieved a nearly level beam, you can make final adjustments by moving the bolt in or out. Adjust the bolt so you have a perfectly level beam.

(13) The final construction step is to make the sliding weights. China uses different standards of weights and measures. You will find weights (actually mass) marked in grams on most products. In the market, the 两 liáng and 斤 jīn are commonly used. The modern day equivalent of these units is that the 两 liáng is equal to 50 g and the 斤 jīn is equal to

Figure 6.40. Preparing a 50.0 g sliding weight using fishing weights.

0.50 kg (approx. 1.1 pounds). You will need to make sliding weights of 1.00 g, 10.0 g, 50.0 g and 1.00×10^2 g. We use different sliding weights depending on the weight range of the object (figure 6.40).

Combine 1 oz, 1/2 oz weights with some split shot (1/10 oz) in order to make 50 g. When you get the mass close to 50 g, use a small length of wire to make up the difference. Use enough wire so that your total mass is slightly more than 50 g. Then use diagonal cutters to snip off small bits of wire. You may even need to file off some of the wire (not the lead) until you obtain a mass of 50.00 g or as close to that as possible. Just file a little and then check the weight. If you need to file the lead weights, be sure to catch the filings in a bowl or some type of container. Lead is hazardous and needs to be disposed of properly. Also make 1.00 g, 10.0 g and 1.00×10^2 g sliding weights. For the 1.00 g weight use the higher precision lab balance (0.01 g). Be sure to wash your hands when finished. Now you are ready to use your 秤 chèng.

How to use the 秤 chèng balance

The operating principle behind the 秤 chèng is static equilibrium. As we have already seen in chapter 4, the product of the force and the distance from the fulcrum is called the torque (τ). This is expressed by the equation

$$\tau = r \times F,$$

where r is the distance from the fulcrum and F is the force that is assumed to act perpendicular to the beam as shown below in figure 6.41. Force F_1 produces a counterclockwise rotation, which is a positive torque. Force F_2 produces a clockwise rotation that gives a negative torque.

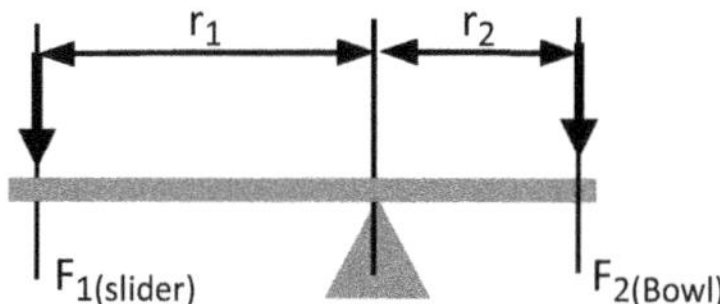

Figure 6.41. Torque model for 秤 chèng.

When the beam is perfectly level, the net sum of the torques is zero. The net torque equation can be written as,

$$\tau_{\text{net}} = (r_1 \times F_1) - (r_2 \times F_2) = 0.$$

Now suppose the force F_2 represents the weight of the object in the bowl (W_b) and F_1 represents the weight of the slider (W_s).

The torque equation becomes

$$r_1 \times W_s = r_2 \times W_b.$$

You prepared several sliding weights, so the weight of the slider W_s is known and the weight in the bowl W_b is the unknown we are trying to determine. Actually, we know the mass of the slider (M_s) and we want to find the mass of the object in the bowl (M_b). Remember that weight is just the gravitational force acting on a mass. So we can substitute

$$W_b = M_b g, \text{ and } W_s = M_s g,$$

where g is the acceleration due to gravity $g = 9.80 \text{ m s}^{-2}$. The torque equation now becomes

$$r_1 \times M_s g = r_2 \times M_b g.$$

The common factor of g on both sides of the equation will divide out and we can solve the equation for the mass of the bowl (M_b). This yields,

$$M_b = \left(\frac{r_1}{r_2}\right) M_s,$$

where r_1 is the distance of the slider from the fulcrum and r_2 is the distance of the bowl from the fulcrum. When we made the notch for the bowl string, we carefully placed it at a distance of 1.00 inches from the fulcrum. Now we can see why that step is so important. With $r_2 = 1.00$ in the equation further simplifies to

$$M_b = r_1 M_s.$$

Suppose you have a 50.0 g slider and the beam balances with the slider at 4.00 inches. In this case the mass is just

$$50.0 \text{ g} \times 4.00 = 2.00 \times 10^2 \text{ g}.$$

Notice that in the step above the units of length divide out in the ratio $(\frac{r_1}{r_2})$.

What if the slider has to be placed at 1 5/8 in. to achieve a level balance? The mass would then be

$$50.0\ \text{g} \times 1\ 5/8 = 81.25\ \text{g, rounded up to } 81.3\ \text{g.}$$

On an actual Chinese 秤 chèng there is a series of markings on the beam and each mark represents an increment of weight. Our beam is calibrated in distance units not mass or weight units so we have to do these simple calculations.

Analysis

Now that we have a working 秤 chèng we need to test it. Like all the experiments described in this book we first duplicate some type of Chinese technology and then subject it to analysis. In the objective we mentioned exploring the difference between systematic and random error. Suppose your watch runs slow compared to some objective time standard such as the National Institute of Standards and Technology (NIST). Every measurement you make will differ from the standard and those differences will always be in the same direction—slow. This is an example of a systematic error. To discover the systematic error we need to have an objective standard and a way to keep track of our measurements compared to that standard. Remember we mentioned the importance of locating the bowl support notch at 1.00 inches. What if it was at 1.10 inches? Then the ratio in the torque equation would be altered. Looking back at the previous example we used a sliding mass of 50.0 g and found the beam was level with the slider at 4.00 inches. If we substitute those values into the torque equation with the bowl notch at 1.10 inches we would have

$$M_b = \left(\frac{r_1}{r_2}\right)M_s = \left(\frac{4.00\ \text{in.}}{1.10\ \text{in.}}\right)50.0\ \text{g}.$$

This would give us a measured mass of 1.82×10^2 g not 2.00×10^2 g. All of our mass measurements would be lower than the mass determined by some objective standard such as a calibrated laboratory weight. Since all of the results are off in the same direction, that is lower, we say there is a systematic error. We can only identify this problem by making a series of measurements, comparing them to a standard and taking note of the general trend.

Now suppose that we fix the systematic error by cutting a new slot or some other means. We still might be faced by a random error. If each of your lab partners measures the exact same standard laboratory weight, will you obtain the same result? You need to try this as part of the experiment. It is not likely that you will get the same result each time. You might view the bubble level a little differently or not have the string on the slider straight along a division mark on the scale. There are all sorts of reasons for the discrepancy. Sometimes you might obtain a value that is larger than the standard value and sometimes lower. This means that there is some distribution of results centered about a mean value. In the field of probability and statistics there are many mathematical ways of expressing the nature of this distribution but we are not going that deep. It is of course possible to have both random and systematic errors occurring at the same time. In that case your mean value would be shifted away from the standard reference value with a distribution of values about the mean value. We usually call the difference between the mean values and the standard value the accuracy. The variation of the values about the mean is referred to at the precision. Going back to the slow watch example, you can have high precision and poor accuracy. This means that your time measurements have a very narrow distribution about the mean and at the same time they are all far away from the standard. Accuracy and precision are illustrated below in figure 6.42. In this example the standard mass is 20.0 g but the measured values have a mean of over 35.0 g with most of the values in the range between 30 g to 42 g.

In addition to precision and accuracy we also need to consider measurement resolution. You may find that the balance might not be very sensitive to small changes in the position of the slider. You can move the slider left or right a little bit and not even see a change in the level. In a way, this sets a fundamental limitation on just how narrow the distribution or precise our measurements can be. You see this clearly when you are trying to make the 1.00 g slider. If you use an electronic mass scale that can only give a reading to 0.1 g resolution can you make a sliding weight to an accuracy of 50.00 g? You have no way of measuring the last digit so the measuring device imposes a limit beyond that of your skill. The ruler scale on the beam is marked in 1/16 inch increments. When you determine the position of the slider you may be able to estimate the value to about half of that distance of 1/32 inch. That gives a fundamental limitation to the mass resolution which will change with different counter weights. In all of this discussion we have talked about the difference between the measured value and the standard value. For the standard value we can use calibrated laboratory weights or we can use an electronic mass scale to measure the mass of an object and then compare the mass determined by the electronic scale to that measured with our 秤 chèng. We can calculate the percentage

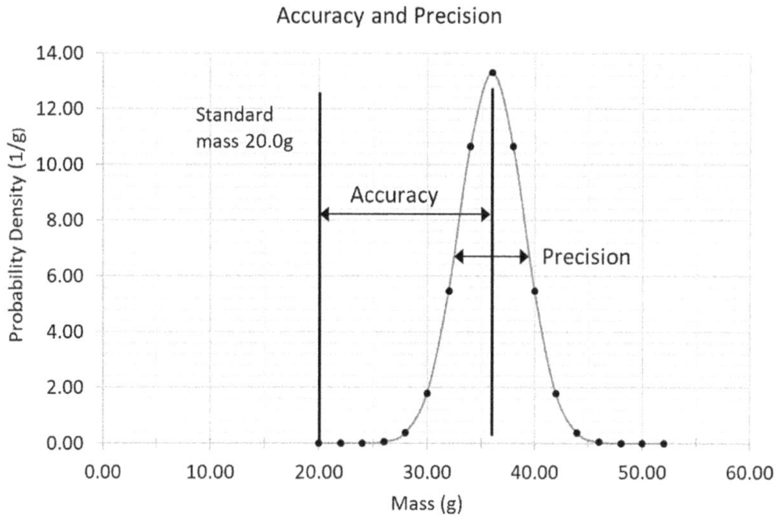

Figure 6.42. Illustration of accuracy and precision showing a systematic error.

difference with the following equation. For the theoretical value we use the mass of the standard weights.

$$\% \text{ difference} = \frac{\text{measured value} - \text{theoretical value}}{\text{theoretical value}} \times 100\%$$

Now we need to characterize the 秤 chèng by evaluating the percentage difference for various mass ranges. We do not expect to have the same accuracy or precision in the 1 g, 100 g and 1 kg ranges. Every measuring instrument has a usable range and we need to determine this. We should also not lose sight of the fact that we are also searching for systematic and random errors. The percentage difference values carry an arithmetic sign with them. If the measured value is larger than the standard values the difference will be positive and if it's smaller the difference will be negative. For random error we expect that some measurements will be a little over the standard value and some will be a little under. On the other hand, systematic errors often reveal themselves by pushing the error to be mostly one sign. We have already addressed one source of systematic error but there are others lurking in the background. You need to consider the possibilities and document them in your laboratory report.

Characterizing the percentage differences

We begin our characterization by selecting standards for comparison. Choose either calibrated laboratory weights or use a calibrated mass scale to measure the mass of several objects. It is always a good idea to check the laboratory weights against a calibrated mass scale. We also need to determine the range of masses we wish to measure with our device. In the past, students have found the 秤 chèng to be usable from about 1 g to 1 kg.

Procedure for characterizing the difference

(1) Select a standard weight and record its mass in the data table.
(2) Record the mass of the slider.
(3) Place the known object into the bowl and record the position of the slider when the beam is balanced.
(4) Calculate the mass and the percentage difference compared to the standard mass.
(5) Repeat this until you have made 15 mass measurements.
(6) For at least one object in each mass range, measure the mass 10 times and record this in the data table. These measurements should be made by different members of the laboratory group. Calculate the mean, standard deviation and coefficient of variation for these measurements.

The standard deviation is given by

$$\sigma = \sqrt{\frac{\sum(x - \bar{x})^2}{(n - 1)}},$$

and the coefficient of variation is $C_v = 100\%\frac{\sigma}{\bar{x}}$.

(7) With the beam balanced, move the slider a little bit to the left and right of the initial value. Record how far the slider must be moved to detect a change in the balance. This determines the measurement resolution. Record these results in your data table and comment on them in your lab report.

(8) Repeat these procedures to cover the high, medium and low ranges from 1 g up to 1 kg. We do not expect the results to be the same over a wide range of masses. You may find that your device has an ideal range where it works best. One way to explore the operating range is to plot a graph of the percentage difference as a function of the mass. Make a graph of the percentage difference as a function of mass for the entire range and graphs for each range. Be sure to observe the sign of the percentage difference. Are there any patterns? Do the differences change sign over the range or are they mostly of the same sign. You may find systematic error in some ranges and random error in other ranges. Comment on this in your lab report. If you were a merchant in ancient times would your customers trust your weight measurements?

Data sheets: For 秤 Chèng steelyard balance experiments

秤 chèng % Difference Characterization- High Range (100g to 1kg) Date____________

Group Names____

Mass of slider __________ Detectable motion of slider Left______ Right______

Standard mass (g)	Slider position	Calculated mass (g)	% Difference

Measure the same object 5 times

Standard Mass	Measured Mass

Average________ STDEV_________ C_v % _________

秤 chèng % Difference Characterization- Medium Range (10g to 100g) Date___________

Group Names____

Mass of slider _________ Detectable motion of slider Left_____ Right_____

Standard mass (g)	Slider position	Calculated mass (g)	% Difference

Measure the same object 5 times

Standard Mass	Measured Mass

Average_________ STDEV_________ C_v % _________

秤 chèng % Difference Characterization- Low Range (1g to 100g) Date___________

Group Names____

Mass of slider _________ Detectable motion of slider Left______ Right______

Standard mass (g)	Slider position	Calculated mass (g)	% Difference

Measure the same object 5 times

Standard Mass	Measured Mass

Average_________ STDEV_________ C_v % _________

References

ASTM Standard A36 2014 *Standard Specification for Carbon Structural Steel* (West Conshohocken, PA: ASTM International)

Barnard N 1975 *The First Radiocarbon Dates from China. Monographs on Far Eastern History xvii* p 94

Kuhn D 1984 Tracing a Chinese legend: in search of the identity of the "First Sericulturalist" *T'oung Pao* **70** 213–45

Needham J, Wang L and Robinson K 1962 *Science and Civilisation in China: 24 Physics and Physical Technology* vol 4 (Cambridge: Cambridge University Press)

Roth H 1912 Oriental steelyards and bismars *J. R. Anthropol. Inst. Great Br. Ireland* **42** 200–33

Zhao H-P, Feng X-Q and Shi H-J 2007 Variability in mechanical properties of Bombyx Mori Silk *Mater. Sci. Eng.* C **27** 675–83

Lightning Source UK Ltd.
Milton Keynes UK
UKHW032229040521
383126UK00007B/529